Introduction to
Modern Genetics

Introduction to Modern Genetics

Dr. J.S. Bohra

Editor

KOROS PRESS LIMITED
London, UK

Introduction to Modern Genetics

© 2012
Printed in 2017 for Sale in the Indian Subcontinent

Published by
Koros Press Limited
3 The Pines, Rubery B45 9FF, Rednal,
Birmingham, United Kingdom

Tel.: +44-7826-930152
Email: info@korospress.com
www.korospress.com

ISBN: 978-1-78163-031-0

Editor: Dr. J.S. Bohra

Printed in UK

British Library Cataloguing in Publication Data
A CIP record for this book is available from the British Library

10 9 8 7 6 5 4 3 2 1

Exclusively distributed by CBS Publishers & Distributors Pvt. Ltd.
Sales & Distribution Rights only for India, Pakistan, Bangladesh, Sri Lanka, Nepal and Bhutan.This book is not to be sold outside these territories.

Contents

Preface   *vii*

1. Somatic Cell Nuclear Transfer   1

Stem Cell Controversy · Stated Views of Groups · Recombinant
DNA · Tissue Engineering · Use of Biotechnology in
Pharmaceutical Manufacturing · Tissue Culture · Cell Culture
· Transfection · RNA Transfection · Transformation (Genetics)
· Established Human Cell Lines · Microbiological Culture
· Organ Culture · Bioeconomics · Biopolitics · Genetic
Engineering · Gene Targeting

2. Comparative Genomics   53

Modelling Biological Systems · Cellular Model · Protein
Structure Prediction · Chou-Fasman Method · GOR Method
· De Novo Protein Structure Prediction · Homology Modelling
· Threading (Protein Sequence) · Protein–protein Interaction
Prediction · CASP · Human Biological Systems · Simulated
Growth of Plants · Ecosystem Model · Mathematical Modelling
of Infectious Disease · Basic Reproduction Number · Epidemic
Model · Macromolecular Docking · Scoring Functions for Docking
· Critical Assessment of Prediction of Interactions
· Biopharmaceutical · Pharming (Genetics) · Molecular Farming

3. Human Genome Project   117

Cloning · Molecular Cloning · Unicellular Organisms · Somatic
Cell Nuclear Transfer · Dolly (Sheep) · Human Cloning
· Ethics of Cloning · Religious Views · Genetically Modified
Food · Genetically Modified Food Controversies · Health Risks
and Benefits · Legal Issues in the US · Controversial Cases
· Biological Engineering · Microbial Biodegradation

4. Molecular Phylogenetics   193

History of Molecular Evolution · Molecular Medicine · Protein
Structure · Amino Acid · Proteinogenic Amino Acid · Protein
Primary Structure · Glycosylation · Deamidation · Hydroxylation
· Methylation · Acetylation · Nucleic Acid Sequence · Sequence
Analysis · Sequence Alignment · Biomolecular Structure
· Expanded Genetic Code · Asymmetric Synthesis · Biodegradable
Plastic · Biopolymer · Peptide Synthesis

Bibliography   275

Index   279

Preface

Comparative genomics exploits both similarities and differences in the proteins, RNA, and regulatory regions of different organisms to infer how selection has acted upon these elements. Those elements that are responsible for similarities between different species should be conserved through time, while those elements responsible for differences among species should be divergent. Finally, those elements that are unimportant to the evolutionary success of the organism will be unconserved. One of the important goals of the field is the identification of the mechanisms of eukaryotic genome evolution. It is however often complicated by the multiplicity of events that have taken place throughout the history of individual lineages, leaving only distorted and superimposed traces in the genome of each living organism. For this reason comparative genomics studies of small model organisms are of great importance to advance our understanding of general mechanisms of evolution. Having come a long way from its initial use of finding functional proteins, comparative genomics is now concentrating on finding regulatory regions and siRNA molecules. Recently, it has been discovered that distantly related species often share long conserved stretches of DNA that do not appear to code for any protein. One such ultra-conserved region, that was stable from chicken to chimp has undergone a sudden burst of change in the human lineage, and is found to be active in the developing brain of the human embryo.

Computational approaches to genome comparison have recently become a common research topic in computer science. A public collection of case studies and demonstrations is growing, ranging from whole genome comparisons to gene expression analysis. This has increased the introduction of different ideas, including concepts from systems and control, information theory, strings analysis and data mining. It is anticipated that computational approaches will become and remain a standard topic for research and teaching, while multiple courses will begin training students to be fluent in both topics. Modelling biological systems is a significant task of systems biology and mathematical biology. Computational systems biology aims to develop and use efficient algorithms, data structures, visualization and communication tools to orchestrate the integration of large quantities of biological data with the goal of computer modelling. It involves the use of computer simulations of biological systems, like cellular subsystems to both analyse and visualize the complex connections of these cellular processes. It is also directly associated with bioinformatics and computational biology. Artificial life or virtual evolution attempts to understand evolutionary processes via the computer simulation of simple life forms.

This book can serve as a basic book for students of molecular biology, genetics, biochemistry, agriculture and biotechnology, or as a reference book for those interested in learning the fundamentals of cell biology.

—Editor

1

Somatic Cell Nuclear Transfer

In genetics and developmental biology, somatic cell nuclear transfer (SCNT) is a laboratory technique for creating a clonal embryo, using an ovum with a donor nucleus. It can be used in embryonic stem cell research, or, potentially, in regenerative medicine where it is sometimes referred to as "therapeutic cloning." It can also be used as the first step in the process of reproductive cloning.

The Process

In SCNT the nucleus, which contains the organism's DNA, of a somatic cell (a body cell other than a sperm or egg cell) is removed and the rest of the cell discarded. At the same time, the nucleus of an egg cell is removed. The nucleus of the somatic cell is then inserted into the denucleated egg cell. After being inserted into the egg, the somatic cell nucleus is reprogrammed by the host cell. The egg, now containing the nucleus of a somatic cell, is stimulated with a shock and will begin to divide. After many mitotic divisions in culture, this single cell forms a blastocyst (an early stage embryo with about 100 cells) with almost identical DNA to the original organism. The technique of transferring a nucleus from a somatic cell into an egg that produced Dolly was an extension of experiments that had been ongoing for over 40 years. In the simplest terms, the technique used to produce Dolly the sheep-somatic cell nuclear transplantation cloning-involves removing the nucleus of an egg and replacing it with the diploid nucleus of a somatic cell.

SCNT in Stem Cell Research

Some researchers use SCNT in stem cell research. The aim of carrying out this procedure is to obtain stem cells that are genetically matched to the donor organism. Presently, no human stem cell lines have been derived from SCNT research. Embryonic stem cells are new, unspecialized cells that are able to be produced into a specialized cell that can replace another cell that has been lost in the body.

A potential use of genetically-customized stem cells would be to create cell lines that have genes linked to the particular disease. For example, if a person with Parkinson's disease donated his or her somatic cells, then the stem cells resulting from SCNT would have genes that contribute to Parkinson's disease. In this scenario, the disease specific stem cell lines would be studied in order to better understand the disease. In another scenario, genetically-customized stem cell lines would be generated for cell-based therapies to transplant to the patient. The resulting cells would be genetically identical to the somatic cell donor, thus avoiding any complications from immune system rejection.

Only a handful of the labs in the world are currently using SCNT techniques in human stem cell research. In the United States, scientists at the Harvard Stem Cell Institute, the University of California San Francisco, Stemagen (La Jolla, CA) and possibly Advanced Cell Technology are currently researching a technique to use somatic cell nuclear transfer to produce embryonic stem cells. In the United Kingdom, the Human Fertilisation and Embryology Authority has granted permission to research groups at the Roslin Institute and the Newcastle Centre for Life. SCNT may also be occurring in China.

In 2005, a South Korean research team led by Professor Hwang Woo-suk, published claims to have derived stem cell lines via SCNT, but supported those claims with fabricated data. Recent evidence has proved that he in fact created a stem cell line from a parthenote.

The impetus for SCNT-based stem cell research has been decreased by the development and improvement of alternative methods of generating stem cells. Methods to reprogram normal body cells into pluripotent stem cells were developed in humans in 2007. The following year, this method achieved a key goal of SCNT-based stem cell research: the derivation of pluripotent stem cell lines that have all genes linked to various diseases. Some scientists working on SCNT-based stem cell research have recently moved to the new methods of induced pluripotent stem cells.

SCNT in Reproductive Cloning

This technique is currently the basis for cloning animals (such as the famous Dolly the sheep), and in theory could be used to clone humans. However, most researchers believe that in the foreseeable future it will not be possible to use this technique to produce a human clone that will develop to term. However, it is still a possibility and can become more probable in the future as it will probably need a few more adjustments to work for humans.

Limitations

Stresses placed on both the egg cell and the introduced nucleus are enormous, leading to a high loss in resulting cells. For example, Dolly the sheep was born after 277 eggs were used for SCNT, which created 29 viable embryos. Only three of these embryos survived until birth, and only one survived to adulthood. As the procedure currently cannot be automated, but has to be performed manually under a microscope, SCNT is very resource intensive.

The biochemistry involved in reprogramming the differentiated somatic cell nucleus and activating the recipient egg is also far from understood. In SCNT, not all of the donor cell's genetic information is transferred, as the donor cell's mitochondria that contain their own mitochondrial DNA are left behind. The resulting hybrid cells retain those mitochondrial structures which originally belonged to the egg. As a consequence, clones such as Dolly that are born from SCNT are not perfect copies of the donor of the nucleus.

Stem Cell Controversy

The stem cell controversy is the ethical debate centered only on research involving the creation, usage, and destruction of human embryos. Most commonly, this controversy focuses on embryonic stem cells. Not all stem cell research involves the creation, usage and destruction of human embryos. For example, adult stem cells or induced pluripotent stem cells, which do not involve creating, using or destroying human embryos, is minimally controversial.

Background

Since stem cells have the ability to differentiate into any type of cell, they offer something in the development of medical treatments for a wide range of conditions. Treatments that have been proposed include treatment for physical trauma, degenerative conditions, and genetic diseases (in combination with gene therapy). Yet further treatments using stem cells could potentially be developed thanks to their ability to repair extensive tissue damage.

Great levels of success and potential have been shown from research using adult stem cells. In early 2009, the FDA approved the first human clinical trials using embryonic stem cells. Some researchers are of the opinion that the differentiation potential of embryonic stem cells is broader than most adult stem cells. Embryonic stem cells can become all cell types of the body because they are pluripotent. Adult stem cells are generally limited to differentiating into different cell types of their tissue of origin. However, some evidence suggests that adult stem cell plasticity may exist, increasing the number of cell types a given adult stem cell can become. In addition, embryonic stem cells are considered more useful for nervous system therapies, because researchers have struggled to identify and isolate neural progenitors from adult tissues. Embryonic stem cells, however, might be rejected by the immune system- a problem which wouldn't occur if the patient received his or her own stem cells.

Some stem cell researchers are working to develop techniques of isolating stem cells that are as potent as embryonic stem cells, but do not require a human embryo.

Some believe that human skin cells can be coaxed to "de-differentiate" and revert to an embryonic state. Researchers at Harvard University, led by Kevin Eggan, have attempted to transfer the nucleus of a somatic cell into an existing embryonic stem cell, thus creating a new stem cell line. Another study published in August 2006 also indicates that differentiated cells can be reprogrammed to an embryonic-like state by introducing four specific factors, resulting in induced pluripotent stem cells.

Researchers at Advanced Cell Technology, led by Robert Lanza, reported the successful derivation of a stem cell line using a process similar to preimplantation genetic diagnosis, in which a single blastomere is extracted from a blastocyst. At the 2007 meeting of the International Society for Stem Cell Research (ISSCR), Lanza announced that his team had succeeded in producing three new stem cell lines without destroying the parent embryos. "These are the first human embryonic cell lines in existence that didn't result from the destruction of an embryo." Lanza is currently in discussions with the National Institutes of Health (NIH) to determine whether the new technique sidesteps U.S. restrictions on federal funding for ES cell research.

Anthony Atala of Wake Forest University says that the fluid surrounding the fetus has been found to contain stem cells that, when utilized correctly, "can be differentiated towards cell types such as fat, bone, muscle, blood vessel, nerve and liver cells". The extraction of this fluid is not thought to harm the fetus in any way. He hopes "that these cells will provide a valuable resource for tissue repair and for engineered organs as well".

Viewpoints

The status of the human embryo and human embryonic stem cell research is a controversial issue as, with the present state of technology, the creation of a human embryonic stem cell line requires the destruction of a human embryo. Stem cell debates have motivated and reinvigorated the pro-life movement, whose members are concerned with the rights and status of the embryo as an early-aged human life. They believe that embryonic stem cell research instrumentalizes and violates the sanctity of life and is tantamount to murder. The fundamental assertion of those who oppose embryonic stem cell research is the belief that human life is inviolable, combined with the fact that human life begins when a sperm cell fertilizes an egg cell to form a single cell.

A portion of stem cell researchers use embryos that were created but not used in vitro fertility treatments to derive new stem cell lines. Most of these embryos are to be destroyed, or stored for long periods of time, long past their viable storage life. In the United States alone, there have been estimates of at least 400,000 such embryos. This has led some opponents of abortion, such as Senator Orrin Hatch, to support human embryonic stem cell research.

Medical researchers widely submit that stem cell research has the potential to dramatically alter approaches to understanding and treating diseases, and to alleviate suffering. In the future, most medical researchers anticipate being able to use technologies derived from stem cell research to treat a variety of diseases and impairments. Spinal cord injuries and Parkinson's disease are two examples that have been championed by high-profile media personalities (for instance, Christopher Reeve and Michael J. Fox, who have lived with these conditions, respectively). The anticipated medical benefits of stem cell research add urgency to the debates, which has been appealed to by proponents of embryonic stem cell research.

In August, 2000, The U.S. National Institutes of Health's Guidelines stated:

> ...*research involving human pluripotent stem cells...promises new treatments and possible cures for*

many debilitating diseases and injuries, including Parkinson s disease, diabetes, heart disease, multiple sclerosis, burns and spinal cord injuries. The NIH believes the potential medical benefits of human pluripotent stem cell technology are compelling and worthy of pursuit in accordance with appropriate ethical standards.f

In 2006, researchers at Advanced Cell Technology of Worcester, Mass., succeeded in obtaining stem cells from mouse embryos without destroying the embryos. If this technique and its reliability are improved, it would alleviate some of the ethical concerns related to embryonic stem cell research.

Another technique announced in 2007 may also defuse the longstanding debate and controversy. Research teams in the United States and Japan have developed a simple and cost effective method of reprogramming human skin cells to function much like embryonic stem cells by introducing artificial viruses.

While extracting and cloning stem cells is complex and extremely expensive, the newly discovered method of reprogramming cells is much cheaper. However, the technique may disrupt the DNA in the new stem cells, resulting in damaged and cancerous tissue. More research will be required before non-cancerous stem cells can be created.

Update article to include 2009/2010 current stem cell usages in clinical trials. The planned treatment trials will focus on the effects of oral lithium on neurological function in people with chronic spinal cord injury and those that have received umbilical cord blood mononuclear cell transplants to the spinal cord. The interest in these two treatments derives from recent reports indicating that umbilical cord blood stem cells may be beneficial for spinal cord injury and that lithium may promote regeneration and recovery of function after spinal cord injury. Both lithium and umbilical cord blood are widely available therapies that have long been used to treat diseases in humans.

Endorsement

- Embryonic stem cells have the potential to grow indefinitely in a laboratory environment and can differentiate into almost all types of bodily tissue. This makes embryonic stem cells a prospect for cellular therapies to treat a wide range of diseases.

Human Potential and Humanity

This argument often goes hand-in-hand with the utilitarian argument, and can be presented in several forms:

- Embryos are not equivalent to human life while they are still incapable of surviving outside the womb (i.e. they only have the potential for life).

- More than a third of zygotes do not implant after conception. Thus, far more embryos are lost due to chance than are proposed to be used for embryonic stem cell research or treatments.

- Blastocysts are a cluster of human cells that have not differentiated into distinct organ tissue; making cells of the inner cell mass no more "human" than a skin cell.

- Some parties contend that embryos are not humans, believing that the life of *Homo sapiens* only begins when the heartbeat develops, which is during the 5th week of pregnancy, or when the brain begins developing activity, which has been detected at 54 days after conception.

Efficiency

- In vitro fertilization (IVF) generates large numbers of unused embryos (e.g. 70,000 in Australia alone). Many of these thousands of IVF embryos are slated for destruction. Using them for scientific research utilizes a resource that would otherwise be wasted.

- While the destruction of human embryos is required to establish a stem cell line, no new embryos have to be destroyed to work with existing stem cell lines. It would be wasteful not to continue to make use of these cell lines as a resource.

- Abortions are legal in many countries and jurisdictions. The argument then follows that if these embryos are being destroyed anyway, why not use them for stem cell research or treatments?

Superiority

This is usually presented as a counter-argument to using adult stem cells as an alternative that doesn't involve embryonic destruction.

- Embryonic stem cells make up a significant proportion of a developing embryo, while adult stem cells exist as minor populations within a mature individual (e.g. in every 1,000 cells of the bone marrow, only 1 will be a usable stem cell).

Thus, embryonic stem cells are likely to be easier to isolate and grow ex vivo than adult stem cells.

- Embryonic stem cells divide more rapidly than adult stem cells, potentially making it easier to generate large numbers of cells for therapeutic means. In contrast, adult stem cell might not divide fast enough to offer immediate treatment.
- Embryonic stem cells have greater plasticity, potentially allowing them to treat a wider range of diseases.
- Adult stem cells from the patient's own body might not be effective in treatment of genetic disorders. Allogeneic embryonic stem cell transplantation (i.e. from a healthy donor) may be more practical in these cases than gene therapy of a patient's own cell.
- DNA abnormalities found in adult stem cells that are caused by toxins and sunlight may make them poorly suited for treatment.
- Embryonic stem cells have been shown to be effective in treating heart damage in mice.
- Embryonic stem cells have the potential to cure chronic and degenerative diseases which current medicine has been unable to effectively treat.

Individuality

- Before the primitive streak is formed when the embryo attaches to the uterus at approximately 14 days after fertilization, a single fertilized egg can split in two to form identical twins, or a pair of embryos that would have resulted in fraternal twins can fuse together and develop into one person (a tetragametic chimera). Since a fertilized egg has the potential to be two individuals or half of one, some believe it can only be considered a *potential* person, not an actual one. Those who subscribe to this belief then hold that destroying a blastocyst for embryonic stem cells is ethical.

Viability

- Viability is another standard under which embryos and fetuses have been regarded as human lives. In the United States, the 1973 Supreme Court case of Roe v. Wade concluded that viability determined the permissibility of abortions performed for reasons other than the protection of the woman's health, defining *viability* as the point at which a fetus is "potentially able to live outside the mother's womb, albeit with artificial aid." The point of viability was 24 to 28 weeks when the case was decided and has since moved to about 22 weeks due to advancement in medical technology. Embryos used in medical research for stem cells are well below below development that would enable viability.

Objection

Value of Life

Embryonic stem cell research currently requires the fertilization of an embryo with the plan to use it as a source for collecting embryonic stem cells, which results in the destruction of the embryo. This deliberate creation and destruction of a human embryo is a primary source for the ethical controversy.

Better Alternatives

This argument is used by opponents of embryonic destruction as well as researchers specializing in adult stem cell research. Pro-life supporters often claim that the use of adult stem cells from sources such as umbilical cord blood has consistently produced more promising results than the use of embryonic stem cells. Furthermore, adult stem cell research may be able to make greater advances if less money and resources were channeled into embryonic stem cell research.

Adult stem cells have already produced therapies, (to date, embryonic stem cells have also been used in treatment) Moreover, there have been many advances in adult stem cell research, including a recent study where pluripotent adult stem cells were manufactured from differentiated fibroblast by the addition of specific transcription factors. Newly created stem cells were developed into an embryo and were integrated into newborn mouse tissues, analogous to the properties of embryonic stem cells. This argument remains hotly debated on both sides. Those critical of embryonic stem cell research point to a current lack of practical treatments, while supporters argue that advances will come with more time and that breakthroughs cannot be predicted.

Using Adult Stem Cells, a possible cure for HIV has been found.

Stated Views of Groups

Government Policy Stances

Europe : Austria, Denmark, France, Germany, and Ireland do not allow the production of embryonic stem cell lines, but the creation of embryonic stem cell lines is permitted in Finland, Greece, the Netherlands, Sweden, and the United Kingdom.

United States : In 1973, Roe v. Wade legalized abortion in the United States. Five years later, the first successful human *in vitro* fertilization resulted in the birth of Louise Brown in England. These developments prompted the federal government to create regulations barring the use of federal funds for research that experimented on human embryos. In 1995, the NIH Human Embryo Research Panel advised the administration of President Bill Clinton to permit federal funding for research on embryos left over from *in vitro* fertility treatments and also recommended federal funding of research on embryos specifically created for experimentation. In response to the panel's recommendations, the Clinton administration, citing moral and ethical concerns, declined to fund research on embryos created solely for research purposes, but did agree to fund research on leftover embryos created by *in vitro* fertility treatments. At this point, the Congress intervened and passed the Dickey Amendment in 1995 (the final bill, which included the Dickey Amendment, was signed into law by Bill Clinton) which prohibited any federal funding for the Department of Health and Human Services be used for research that resulted in the destruction of an embryo regardless of the source of that embryo.

In 1998, privately funded research led to the breakthrough discovery of Human Embryonic Stem Cells (hESC). This prompted the Clinton Administration to re-examine guidelines for federal funding of embryonic research. In 1999, the president's National Bioethics Advisory Commission recommended that hESC harvested from embryos discarded after *in vitro* fertility treatments, but not from embryos created expressly for experimentation, be eligible for federal funding. Though embryo destruction had been inevitable in the process of harvesting hESC in the past (this is no longer the case), the Clinton Administration had decided that it would be permissible under the Dickey Amendment to fund hESC research as long as such research did not itself directly cause the destruction of an embryo. Therefore, HHS issued its proposed regulation concerning hESC funding in 2001. Enactment of the new guidelines was delayed by the incoming George W. Bush administration which decided to reconsider the issue.

President Bush announced, on August 9, 2001 that federal funds, for the first time, would be made available for hESC research on currently existing embryonic stem cell lines. President Bush authorized research on existing human embryonic stem cell lines, not on human embryos under a specific, unrealistic timeline in which the stem cell lines must have been developed. However, the Bush Administration chose not to permit taxpayer funding for research on hESC cell lines not currently in existence, thus limiting federal funding to research in which "the life-and-death decision has already been made". The Bush Administration's guidelines differ from the Clinton Administration guidelines which did not distinguish between currently existing and not-yet-existing hESC. Both the Bush and Clinton guidelines agree that the federal government should not fund hESC research that directly destroys embryos.

Neither Congress nor any administration has ever prohibited private funding of embryonic research. Public and private funding of research on adult and cord blood stem cells is unrestricted.

U.S. Congressional Response

In April 2004, 206 members of Congress signed a letter urging President Bush to expand federal funding of embryonic stem cell research beyond what Bush had already supported.

In May 2005, the House of Representatives voted 238-194 to loosen the limitations on federally funded embryonic stem-cell research — by allowing government-funded research on surplus frozen embryos from in vitro fertilization clinics to be used for stem cell research with the permission of donors — despite Bush's promise to veto the bill if passed. On July 29, 2005, Senate Majority Leader William H. Frist (R-TN), announced that he too favoured loosening restrictions on federal funding of embryonic stem cell research. On July 18, 2006, the Senate passed three different bills concerning stem cell research. The Senate passed the first bill (Stem Cell Research Enhancement Act), 63-37, which would have made it legal for the Federal government to spend Federal money on embryonic stem cell research that uses embryos left over from *in vitro* fertilization procedures. On July 19, 2006 President Bush vetoed this bill. The second bill makes it illegal to create, grow, and abort fetuses for research purposes. The third bill would encourage research that would isolate pluripotent, i.e., embryonic-like, stem cells without the destruction of human embryos.

In 2005 and 2007, Congressman Ron Paul introduced the Cures Can Be Found Act, with 10 cosponsors. With an income tax credit, the bill favours research upon non embryonic stem cells obtained from placentas, umbilical cord blood, amniotic fluid, humans after birth, or unborn human offspring who died of natural causes; the bill was referred to committee. Paul argued that hESC research is outside of federal jurisdiction either to ban or to subsidize.

Bush vetoed another bill, the Stem Cell Research Enhancement Act of 2007, which would have amended the Public Health Service Act to provide for human embryonic stem cell research. The bill passed the Senate on April 11 by a vote of 63-34, then passed the House on June 7 by a vote of 247-176. President Bush vetoed the bill on July 19, 2007.

On March 9, 2009, President Obama removed the restriction on federal funding for newer stem cell lines. Two days after Obama removed the restriction, the President then signed the Omnibus Appropriations Act of 2009, which still contained the long-standing Dickey-Wicker provision which bans federal funding of "research in which a human embryo or embryos are destroyed, discarded, or knowingly subjected to risk of injury or death;" the Congressional provision effectively prevents federal funding being used to create new stem cell lines by many of the known methods. So, while scientists might not be free to create new lines with federal funding, President Obama's policy allows the potential of applying for such funding into research involving the hundreds of existing stem cell lines as well as any further lines created using private funds or state-level funding. The ability to apply for federal funding for stem cell lines created in the private sector is a significant expansion of options over the limits imposed by President Bush, who restricted funding to the 21 viable stem cell lines that were created before he announced his decision in 2001. The ethical concerns raised during Clinton's time in office continue to restrict hESC research and dozens of stem cell lines have been excluded from funding, now by judgment of an administrative office rather than Presidential or legislative discretion.

Funding

In 2005 the NIH funded $607 million worth of stem cell research, of which $39 million was specifically used for hESC. Sigrid Fry-Revere has argued that private organizations, not the federal government, should provide funding for stem-cell research, so that shifts in public opinion and government policy would not bring valuable scientific research to a grinding halt.

In 2005 the State of California took out 3 billion dollars in bond loans to fund embryonic stem cell research in that state.

Church Views

Baptists: The Southern Baptist Convention opposes human embryonic stem cell research on the grounds that "Bible teaches that human beings are made in the image and likeness of God (Gen. 1:27; 9:6) and protectable human life begins at fertilization." However, it supports adult stem cell research as it does "not require the destruction of embryos."

Catholicism

In regards, to embryonic stem cell research, the Catholic Church affirms that "the killing of innocent human creatures, even if carried out to help others, constitutes an absolutely unacceptable act." The deliberate destruction of a human embryo is incompatible with Roman Catholic doctrine, according to which, Pontifical Academy for Life has stated that human blastocysts are inherently valuable and should not be voluntarily destroyed as they are "from the moment of the union of the gametes" human subjects with well defined identities.

The Church, however, supports research that involves stem cells from adult tissues and the umbilical cord, as it "involves no harm to human beings at any state of development."

Methodism

In regards, to embryonic stem cell research, the United Methodist Church stands in "opposition to the creation of embryos for the sake of research" as "a human embryo, even at its earliest stages, commands our reverence." However, it supports adult stem cell research, stating that there are "few moral questions" raised by this issue.

Pentecostalism

The Assemblies of God opposes the "the practice of cultivating stem cells from the tissue of aborted fetuses" because it "perpetuates the evil of abortion and should be prohibited."

Jewish View

According to Rabbi Levi Yitschak Halperin of the Institute for Science and Jewish Law in Jerusalem, embryonic stem cell research is permitted so long as it has not been implanted in the womb. Not only is it permitted, but research is encouraged, rather than wasting it.

> *As long as it has not been implanted in the womb and it is still a frozen fertilized egg, it does not have the status of an embryo at all and there is no prohibition*

to destroy it. However in order to remove all doubt [as to the permissibility of destroying it], it is preferable not to destroy the pre-embryo unless it will otherwise not be implanted in the woman who gave the eggs (either because there are many fertilized eggs, or because one of the parties refuses to go on with the procedure- the husband or wife- or for any other reason). Certainly it should not be implanted into another woman.... The best and worthiest solution is to use it for life-saving purposes, such as for the treatment of people that suffered trauma to their nervous system, etc.f

—Rabbi Levi Yitschak Halperin, Ma'aseh Choshev.

Policies Regarding Human SCNT

SCNT involving human cells is currently legal for research purposes in the United Kingdom, having been incorporated into the Human Fertilisation and Embryology Act 1990 in 2001. Permission must be obtained from the Human Fertilisation and Embryology Authority in order to perform or attempt SCNT. In the United States, the practice remains legal, as it has not been addressed by federal law. However, in 2002, a moratorium on United States federal funding for SCNT prohibits funding the practice for the purposes of research.

Thus, though legal, SCNT cannot be federally funded. In 2003, the United Nations adopted a proposal submitted by Costa Rica, calling on member states to "prohibit all forms of human cloning in as much as they are incompatible with human dignity and the protection of human life." This phrase may include SCNT, depending on interpretation. The Council of Europe's *Convention on Human Rights and Biomedicine* and its *Additional Protocol to the Convention for the Protection of Human Rights and Dignity of the Human Being with regard to the Application of Biology and Medicine, on the Prohibition of Cloning Human Being* appear to ban SCNT of human beings. Of the Council's 45 member states, the *Convention* has been signed by 31 and ratified by 18. The *Additional Protocol* has been signed by 29 member nations and ratified by 14.

The UN is currently against all forms of human cloning.

Recombinant DNA

Recombinant DNA (rDNA) is a form of artificial DNA that is created by combining two or more sequences that would not normally occur together through the process of gene splicing. In terms of genetic modification, it is created through the introduction of relevant DNA into an existing organismal DNA, such as the plasmids of bacteria, to code for or alter different traits for a specific purpose, such as antibiotic resistance. It differs from genetic recombination in that it does not occur through natural processes within the cell, but is engineered. A recombinant protein is a protein that is derived from recombinant DNA.

Methods

Cloning and Relation to Plasmids: The use of cloning is interrelated with recombinant DNA in classical biology, as the term "clone" refers to a cell or organism derived from a parental organism, with modern biology referring to the term as a collection of cells derived from the same cell that remain identical. In the classical instance, the use of recombinant DNA provides the initial cell from which the host organism is then expected to recapitulate when it undergoes further cell division, with bacteria remaining a prime example due to the use of viral vectors in medicine that contain recombinant DNA inserted into a structure known as a plasmid.

Plasmids are extrachromosomal self-replicating circular forms of DNA present in most bacteria, such as *Escherichia coli* (E. Coli), containing genes related to catabolism and metabolic activity, and allowing the carrier bacterium to survive and reproduce in conditions present within other species and environments. These genes represent characteristics of resistance to bacteriophages and antibiotics and some heavy metals, but can also be fairly easily removed or separated from the plasmid by restriction endonucleases, which regularly produce "sticky ends" and allow the attachment of a selected segment of DNA, which codes for more "reparative" substances, such as peptide hormone medications including insulin, growth hormone, and oxytocin. In the introduction of useful genes into the plasmid, the bacteria are then used as a viral vector, which are encouraged to reproduce so as to recapitulate the altered DNA within other cells it infects, and increase the amount of cells with the recombinant DNA present within them.

The use of plasmids is also key within gene therapy, where their related viruses are used as cloning vectors or carriers, which are means of transporting and passing on genes in recombinant DNA through viral reproduction throughout an organism. Plasmids contain three common features—a replicator, selectable marker and a cloning site. The replicator or "ori" refers to the origin of replication with regard to location and bacteria where replication begins.

The marker refers to a particular gene that usually contains resistance to an antibiotic, but may also refer to a gene that is attached alongside the desired one, such as that which confers luminescence to allow identification of successfully recombined DNA. The cloning site is a sequence of nucleotides representing one or more positions where cleavage by restriction endonucleases occurs. Most eukaryotes do not maintain canonical plasmids; yeast is a notable exception. In addition, the Ti plasmid of the bacterium *Agrobacterium tumefaciens* can be used to integrate foreign DNA into the genomes of many plants. Other methods of introducing or creating recombinant DNA in eukaryotes include homologous recombination and transfection with modified viruses.

Chimeric Plasmids

When recombinant DNA is then further altered or changed to host additional strands of DNA, the molecule formed is referred to as "chimeric" DNA molecule, with reference to the mythological chimera, which consisted as a composite of several animals. The presence of chimeric plasmid molecules is somewhat regular in occurrence, as, throughout the lifetime of an organism, the propagation by vectors ensures the presence of hundreds of thousands of organismal and bacterial cells that all contain copies of the original chimeric DNA. In the production of chimeric (from chimera) plasmids, the processes involved can be somewhat uncertain, as the intended outcome of the addition of foreign DNA may not always be achieved and may result in the formation of unusable plasmids. Initially, the plasmid structure is linearised to allow the addition by bonding of complementary foreign DNA strands to single-stranded "overhangs" or "sticky ends" present at the ends of the DNA molecule from staggered, or "S-shaped" cleavages produced by restriction endonucleases.

A common vector used for the donation of plasmids originally was the bacterium Escherichia coli and, later, the EcoRI derivative, which was used for its versatility with addition of new DNA by "relaxed" replication when inhibited by chloramphenicol and spectinomycin, later being replaced by the pBR322 plasmid. In the case of EcoRI, the plasmid can anneal with the presence of foreign DNA via the route of sticky-end ligation, or with "blunt ends" via blunt-end ligation, in the presence of the phage T_4 ligase, which forms covalent links between 3-carbon OH and 5-carbon PO_4 groups present on blunt ends. Both sticky-end, or overhang ligation and blunt-end ligation can occur between foreign DNA segments, and cleaved ends of the original plasmid depending upon the restriction endonuclease used for cleavage.

Applications

There are multitudinous proteins that are created from recombinant DNA and used as medications. Some can alternatively be produced from animal extracts or harvested from humans, such as human growth hormone (rhGH), human insulin, follicle-stimulating hormone (FSH) and factor VIII. Other proteins, when used as medication, only has recombinant DNA as a source, such as with erythropoietin.

History

The recombinant DNA technique was first proposed by Peter Lobban, a graduate student, with A. Dale Kaiser at the Stanford University Department of Biochemistry. The technique was then realized by Lobban and Kaiser; Jackson, Symons and Berg; and Stanley Norman Cohen, Chang, Herbert Boyer and Helling, in 1972–74. They published their findings in papers including the 1972 paper *"Biochemical Method for Inserting New Genetic Information into DNA of Simian Virus 40: Circular SV40 DNA Molecules Containing Lambda Phage Genes and the Galactose Operon of Escherichia coli"*, the 1973 paper *"Enzymatic end-to-end joining of DNA molecules"* and the 1973 paper *Construction of Biologically Functional Bacterial Plasmids* in vitro*, all of which described techniques to isolate and amplify genes or DNA segments and insert them into another cell with precision, creating a transgenic bacterium. Recombinant DNA technology was made possible by the discovery, isolation and application of restriction endonucleases by Werner Arber, Daniel Nathans, and Hamilton Smith, for which they received the 1978 Nobel Prize in Medicine. Cohen and Boyer applied for a patent on the Process for producing biologically functional molecular chimeras which could not exist in nature in 1974. The patent was granted in 1980. A breakthrough in recombinant DNA technology occurred in 1977 when Herbert Boyer produced biosynthetic "human" insulin in the lab. The specific gene sequence, or polynucleotide, that codes for insulin production in humans was introduced to a sample colony of the *E. coli* bacteria. It was the first medicine made via recombinant DNA technology to be approved by the FDA and commercially available under the brand name Humulin. The vast majority of insulin currently used worldwide is now biosynthetic recombinant "human" insulin or its analogs.

Tissue Engineering

Tissue engineering was once categorized as a sub-field of bio materials, but having grown in scope and importance it can be considered as a field in its own right. It is the use of a combination of cells, engineering and materials methods, and suitable biochemical and physio-chemical factors to improve or replace biological functions. While most definitions of tissue engineering cover a broad range of applications, in practice the term is closely associated with applications that repair or replace portions of or whole tissues (i.e., bone, cartilage, blood vessels, bladder, skin etc.). Often, the tissues involved require certain mechanical and structural properties for proper functioning. The term has also been applied to efforts to perform specific biochemical functions using cells within an artificially-created support system (e.g. an artificial pancreas, or a bio artificial liver). The term regenerative medicine is often used synonymously with tissue engineering, although those involved in regenerative medicine place more emphasis on the use of stem cells to produce tissues.

Commonly applied definition of tissue engineering, as stated by Langer and Vacanti, is "an interdisciplinary field that applies the principles of engineering and life sciences toward the development of biological substitutes that restore, maintain, or improve tissue function or a whole organ". Tissue engineering has also been defined as "understanding the principles of tissue growth, and applying this to produce functional replacement tissue for clinical use." A further description goes on to say that an "underlying supposition of tissue engineering is that the employment of natural biology of the system will allow for greater success in developing therapeutic strategies aimed at the replacement, repair, maintenance, and/or enhancement of tissue function."

Powerful developments in the multidisciplinary field of tissue engineering have yielded a novel set of tissue replacement parts and implementation strategies. Scientific advances in biomaterials, stem cells, growth and differentiation factors, and biomimetic environments have created unique opportunities to fabricate tissues in the laboratory from combinations of engineered extracellular matrices ("scaffolds"), cells, and biologically active molecules. Among the major challenges now facing tissue engineering is the need for more complex functionality, as well as both functional and biomechanical stability in laboratory-grown tissues destined for transplantation. The continued success of tissue engineering, and the eventual development of true human replacement parts, will grow from the convergence of engineering and basic research advances in tissue, matrix, growth factor, stem cell, and developmental biology, as well as materials science and bio informatics.

In 2003, the NSF published a report entitled "The Emergence of Tissue Engineering as a Research Field", which gives a thorough description of the history of this field.

Use of Biotechnology in Pharmaceutical Manufacturing

Modern pharmaceutical manufacturing techniques frequently rely upon biotechnology.

Human Insulin

Amongst the earliest uses of biotechnology in pharmaceutical manufacturing is the use of recombinant DNA technology to modify *Escherichia coli* bacteria to produce human insulin, which was performed at Genentech in 1978. Prior to the development of this technique, insulin was extracted from the pancreas glands of cattle, pigs, and other farm animals. While generally efficacious in the treatment of diabetes, animal-derived insulin is not indistinguishable from human insulin, and may therefore produce allergic reactions. Genentech researchers produced artificial genes for each of the two protein chains that comprise the insulin molecule. The artificial genes were "then inserted... into plasmids... among a group of genes that" are activated by lactose. Thus, the insulin producing genes were also activated by lactose. The recombinant plasmids were inserted into *Escherichia coli* bacteria, which were "induced to produce 100,000 molecules of either chain A or chain B human insulin." The two protein chains were then combined to produce insulin molecules.

Human Growth Hormone

Prior to the use of recombinant DNA technology to modify bacteria to produce human growth hormone, the hormone was manufactured by extraction from the pituitary glands of cadavers, as animal growth hormones have no therapeutic value in humans. Production of a single year's supply of human growth hormone required up to fifty pituitary glands, creating significant shortages of the hormone. In 1979, scientists at Genentech produced human growth hormone by inserting DNA coding for human growth hormone into a plasmid that was implanted in escherichia coli bacteria. The gene that was inserted into the plasmid was created by reverse transcription of the mRNA found in pituitary glands to complementary DNA. HaeIII, a type of restriction enzyme which acts at restriction sites "in the 3' noncoding region" and at the 23rd codon in complementary DNA for human growth hormone, was used to produce "a DNA fragment of 551 base pairs which includes coding sequences for amino acids 24–191 of HGH." Then "a chemically synthesized DNA 'adaptor' fragment containing an ATG initiation codon..." was produced with the codons for the first through 23rd amino acids in human growth hormone. The "two DNA fragments... [were] combined to form a synthetic-natural 'hybrid' gene." The use of entirely synthetic methods of DNA production to produce a gene that would be translated to human growth hormone in escherichia coli would have been exceedingly laborious due to the significant length of the amino acid sequence in human growth hormone. However, if the cDNA reverse transcribed from the mRNA for human growth hormone were inserted directly into the plasmid inserted into the escherichia coli, the bacteria would translate regions of the gene that are not translated in humans, thereby producing a "pre-hormone containing an extra 26 amino acids" which might be difficult to remove.

Human Blood Clotting Factors

Prior to the development and FDA approval of a means to produce human blood clotting factors using recombinant DNA technologies, human blood clotting factors were produced from donated blood that was inadequately screened for HIV. Thus, HIV infection posed a significant danger to patients with hemophilia who received human blood clotting factors:

Most reports indicate that 60 to 80 percent of patients with hemophilia who were exposed to factor VIII concentrates between 1979 and 1984 are seropositive for HIV by [the] Western blot assay. As of May 1988, more than 659 patients with hemophilia had AIDS...

The first human blood clotting factor to be produced in significant quantities using recombinant DNA technology was Factor IX, which was produced using transgenic Chinese hamster ovary cells in 1986. Lacking a map of the human genome, researchers obtained a known sequence of the RNA for Factor IX by examining the amino acids in Factor IX:

Microsequencing of highly purified... [Factor IX] yielded sufficient amino acid sequence to construct oligonucleotide probes.

The known sequence of Factor IX RNA was then used to search for the gene coding for Factor IX in a library of the DNA found in the human liver, since it was known that blood clotting factors are produced by the human liver:

A unique oligonucleotide... homologous to Factor IX mRNA... was synthesized and labelled... The resultant probe was used to screen a human liver double-stranded cDNA library... Complete two-stranded DNA sequences of the... [relevant] cDNA... contained all of the coding sequence COOH-terminal of the eleventh codon (11) and the entire 3'-untranslated sequence.

This sequence of cDNA was used to find the remaining DNA sequences comprising the Factor IX gene by searching the DNA in the X chromosome:

A genomic library from a human XXXX chromosome was prepared... and screen[ed] with a Factor IX cDNA probe. Hybridizing recombinant phage were isolated, plaque-purified, and the DNA isolated. Restriction mapping, Southern analysis, and DNA sequencing permitted identification of five recombinant phage-containing inserts which, when overlapped at common sequences, coded the entire 35kb Factor IX gene.

Plasmids containing the Factor IX gene, along with plasmids with a gene that codes for resistance to methotrexate, were inserted into Chinese hamster ovary cells via transfection. Transfection involves the insertion of DNA into a eukaryotic cell. Unlike the analogous process of transformation in bacteria, transfected DNA is not ordinarily integrated into the cell's genome, and is therefore not usually passed on to subsequent generations via cell division. Thus, in order to obtain a "stable" transfection, a gene which confers a significant survival advantage must also be transfected, causing the few cells that did integrate the transfected DNA into their genomes to increase their population as cells that did not integrate the DNA are eliminated. In the case of this study, "grow[th] in increasing concentrations of methotrexate" promoted the survival of stably transfected cells, and diminished the survival of other cells.

The Chinese hamster ovary cells that were stably transfected produced significant quantities of Factor IX, which was shown to have substantial coagulant properties, though of a lesser degree than Factor IX produced from human blood: The specific activity of the recombinant Factor IX was measured on the basis of direct measurement of the coagulant activity... The specific activity of recombinant Factor IX was 75 units/mg... compared to 150 units/mg measured for plasma-derived

Factor IX... In 1992, the FDA approved Factor VIII produced using transgenic Chinese hamster ovary cells, the first such blood clotting factor produced using recombinant DNA technology to be approved.

Transgenic Farm Animals

Recombinant DNA techniques have also been employed to create transgenic farm animals that can produce pharmaceutical products for use in humans. For instance, pigs that produce human hemoglobin have been created. While blood from such pigs could not be employed directly for transfusion to humans, the hemoglobin could be refined and employed to manufacture a blood substitute.

Tissue Culture

Tissue culture is the growth of tissues and/or cells separate from the organism. This is typically facilitated via use of a liquid, semi-solid, or solid growth medium, such as broth or agar. Tissue culture commonly refers to the culture of animal cells and tissues, while the more specific term plant tissue culture is being named for the plants.

Historical Usage

In 1885 Wilhelm Roux removed a section of the medullary plate of an embryonic chicken and maintained it in a warm saline solution for several days, establishing the basic principle of tissue culture. In 1907 the zoologist Ross Granville Harrison demonstrated the growth of frog nerve cell processes in a medium of clotted lymph. In 1913, E. Steinhardt, C. Israeli, and R. A. Lambert grew vaccinia virus in fragments of guinea pig corneal tissue.

Cell Culture

Cell culture is the complex process by which cells are grown under controlled conditions. In practice, the term "cell culture" has come to refer to the culturing of cells derived from multicellular eukaryotes, especially animal cells. However, there are also cultures of plants, fungi and microbes, including viruses, bacteria and protists. The historical development and methods of cell culture are closely interrelated to those of tissue culture and organ culture. Animal cell culture became a common laboratory technique in the mid-1900s, but the concept of maintaining live cell lines separated from their original tissue source was discovered in the 19th century.

History

The 19th-century English physiologist Sydney Ringer developed salt solutions containing the chlorides of sodium, potassium, calcium and magnesium suitable for maintaining the beating of an isolated animal heart outside of the body. In 1885 Wilhelm Roux removed a portion of the medullary plate of an embryonic chicken and maintained it in a warm saline solution for several days, establishing the principle of tissue culture. Ross Granville Harrison, working at Johns Hopkins Medical School and then at Yale University, published results of his experiments from 1907–1910, establishing the methodology of tissue culture.

Cell culture techniques were advanced significantly in the 1940s and 1950s to support research in virology. Growing viruses in cell cultures allowed preparation of purified viruses for the manufacture of vaccines. The injectable polio vaccine developed by Jonas Salk was one of the first products mass-produced using cell culture techniques. This vaccine was made possible by the cell culture research of John Franklin Enders, Thomas Huckle Weller, and Frederick Chapman Robbins, who were awarded a Nobel Prize for their discovery of a method of growing the virus in monkey kidney cell cultures.

Concepts in Mammalian Cell Culture

Isolation of Cells: Cells can be isolated from tissues for *ex vivo* culture in several ways. Cells can be easily purified from blood, however only the white cells are capable of growth in culture. Mononuclear cells can be released from soft tissues by *enzymatic digestion* with enzymes such as collagenase, trypsin, or pronase, which break down the extracellular matrix. Alternatively, pieces of tissue can be placed in growth media, and the cells that grow out are available for culture. This method is known as *explant culture*.

Cells that are cultured directly from a subject are known as *primary cells*. With the exception of some derived from tumours, most primary cell cultures have limited lifespan. After a certain number of population doublings (called the Hayflick limit) cells undergo the process of senescence and stop dividing, while generally retaining viability. An established or immortalised *cell line* has acquired the ability to proliferate indefinitely either through random mutation or deliberate modification, such as artificial expression of the telomerase gene. There are numerous well established cell lines representative of particular cell types.

Maintaining Cells in Culture

Cells are grown and maintained at an appropriate temperature and gas mixture (typically, 37°C, 5% CO_2 for mammalian cells) in a cell incubator. Culture conditions vary widely for each cell type, and variation of conditions for a particular cell type can result in different phenotypes being expressed.

Aside from temperature and gas mixture, the most commonly varied factor in culture systems is the growth medium. Recipes for growth media can vary in pH, glucose concentration, growth factors, and the presence of other nutrients. The growth factors used to supplement media are often derived from animal blood, such as calf serum. One complication of these blood-derived ingredients is the potential for contamination of the culture with viruses or prions, particularly in biotechnology medical applications. Current practice is to minimize or eliminate the use of these ingredients wherever possible, but this cannot always be accomplished. Alternative strategies involve sourcing the animal blood from countries with minimum BSE/TSE risk such as Australia and New Zealand, and using purified nutrient concentrates derived from serum in place of whole animal serum for cell culture.

Plating density (number of cells per volume of culture medium) plays a critical role for some cell types. For example, a lower plating density makes granulosa cells exhibit estrogen production, while a higher plating density makes them appear as progesterone producing theca lutein cells. Cells can be grown in *suspension* or *adherent* cultures. Some cells naturally live in suspension, without being attached to a surface, such as cells that exist in the bloodstream. There are also cell lines that have been modified to be able to survive in suspension cultures so that they can be grown to a higher density than adherent conditions would allow.

Adherent cells require a surface, such as tissue culture plastic or microcarrier, which may be coated with extracellular matrix components to increase adhesion properties and provide other signals needed for growth and differentiation. Most cells derived from solid tissues are adherent. Another type of adherent culture is *organotypic culture* which involves growing cells in a three-dimensional environment as opposed to two-dimensional culture dishes. This 3D culture system is biochemically and physiologically more similar to *in vivo* tissue, but is technically challenging to maintain because of many factors (e.g. diffusion).

Cell Line Cross-contamination

Cell line cross-contamination can be a problem for scientists working with cultured cells. Studies suggest that anywhere from 15–

20% of the time, cells used in experiments have been misidentified or contaminated with another cell line. Problems with cell line cross contamination have even been detected in lines from the NCI-60 panel, which are used routinely for drug-screening studies. Major cell line repositories including the American Type Culture Collection (ATCC) and the German Collection of Microorganisms and Cell Cultures (DSMZ) have received cell line submissions from researchers that were misidentified by the researcher. Such contamination poses a problem for the quality of research produced using cell culture lines, and the major repositories are now authenticating all cell line submissions. ATCC uses short tandem repeat (STR) DNA fingerprinting to authenticate its cell lines.

To address this problem of cell line cross-contamination, researchers are encouraged to authenticate their cell lines at an early passage to establish the identity of the cell line. Authentication should be repeated before freezing cell line stocks, every two months during active culturing and before any publication of research data generated using the cell lines. There are many methods for identifying cell lines including isoenzyme analysis, human lymphocyte antigen (HLA) typing, Chromosomal analysis, Karyotyping, Morphology and STR analysis.

One significant cell-line cross contaminant is the immortal HeLa cell line.

Manipulation of Cultured Cells

As cells generally continue to divide in culture, they generally grow to fill the available area or volume. This can generate several issues:

- Nutrient depletion in the growth media
- Accumulation of apoptotic/necrotic (dead) cells.
- Cell-to-cell contact can stimulate cell cycle arrest, causing cells to stop dividing known as contact inhibition or senescence.
- Cell-to-cell contact can stimulate cellular differentiation.

Among the common manipulations carried out on culture cells are media changes, passaging cells, and transfecting cells. These are generally performed using tissue culture methods that rely on sterile technique. Sterile technique aims to avoid contamination with bacteria, yeast, or other cell lines. Manipulations are typically carried out in a biosafety hood or laminar flow cabinet to exclude contaminating micro-organisms. Antibiotics (e.g. penicillin and streptomycin) and antifungals (e.g. Amphotericin B) can also be added to the growth media. As cells undergo metabolic processes, acid is produced and the pH decreases. Often, a pH indicator is added to the medium in order to measure nutrient depletion.

Media Changes

In the case of adherent cultures, the media can be removed directly by aspiration and replaced.

Passaging Cells

Passaging (also known as subculture or splitting cells) involves transferring a small number of cells into a new vessel. Cells can be cultured for a longer time if they are split regularly, as it avoids the senescence associated with prolonged high cell density. Suspension cultures are easily passaged with a small amount of culture containing a few cells diluted in a larger volume of fresh media. For adherent cultures, cells first need to be detached; this is commonly done with a mixture of trypsin-EDTA, however other enzyme mixes are now available for this purpose. A small number of detached cells can then be used to seed a new culture.

Transfection

Transfection is the process of deliberately introducing nucleic acids into cells. The term is used notably for non-viral methods in eukaryotic cells. It may also refer to other methods and cell types, although other terms are preferred: "transformation" is more often used to describe non-viral DNA transfer in bacteria, non-animal eukaryotic cells and plant cells-a distinctive sense of transformation refers to spontaneous genetic modifications (mutations to cancerous cells (Carcinogenesis), or under stress (UV irradiation)). "Transduction" is often used to describe virus-mediated DNA transfer. The word *transfection* is a blend of *trans-* and *infection*.

Genetic material (such as supercoiled plasmid DNA or siRNA constructs), or even proteins such as antibodies, may be transfected.

Transfection of animal cells typically involves opening transient pores or "holes" in the cell membrane, to allow the uptake of material. Transfection can be carried out using calcium phosphate, by electroporation, or by mixing a cationic lipid with the material to produce liposomes, which fuse with the cell membrane and deposit their cargo inside.

Transfection can result in unexpected morphologies and abnormalities in target cells.

Terminology

The meaning of the term has evolved. The original meaning of transfection was "infection by transformation", *i.e.* introduction of DNA (or RNA) from a prokaryote-infecting virus or bacteriophage into cells, resulting in an infection. Because the term transformation had another sense in animal cell biology (a genetic change allowing long-term propagation in culture, or acquisition of properties typical of cancer cells), the term transfection acquired, for animal cells, its present meaning of a change in cell properties caused by introduction of DNA.

Methods

There are various methods of introducing foreign DNA into a eukaryotic cell: some rely on physical treatment (electroporation, nanoparticles, magnetofection), other on chemical materials or biological particles (viruses) that are used as carriers.

Chemical-based Transfection

Chemical-based transfection can be divided into several kinds: cyclodextrin, polymers, liposomes, or nanoparticles (with or without chemical or viral functionalization).

- One of the cheapest methods uses calcium phosphate, originally discovered by F. L. Graham and A. J. van der Eb in 1973. HEPES-buffered saline solution (HeBS) containing phosphate ions is combined with a calcium chloride solution containing the DNA to be transfected. When the two are combined, a fine precipitate of the positively charged calcium and the negatively charged phosphate will form, binding the DNA to be transfected on its surface. The suspension of the precipitate is then added to the cells to be transfected (usually a cell culture grown in a monolayer). By a process not entirely understood, the cells take up some of the precipitate, and with it, the DNA.

- Other methods use highly branched organic compounds, so-called dendrimers, to bind the DNA and get it into the cell.

- A very efficient method is the inclusion of the DNA to be transfected in liposomes, i.e. small, membrane-bounded bodies that are in some ways similar to the structure of a cell and can actually fuse with the cell membrane, releasing the DNA into the cell. For eukaryotic cells, transfection is better achieved using cationic liposomes (or mixtures), because the cells are more sensitive. Popular agents were DOTMA and DOPE, and now- more effectively- Lipofectamine and UptiFectin.

- Another method is the use of cationic polymers such as DEAE-dextran or polyethylenimine. The negatively charged DNA binds to the polycation and the complex is taken up by the cell via endocytosis. Popular agents of this type are the Fugene or LT-1, and JetPEI.

- Other proprietary chemical transfection reagents: PromoFectin, GenePORTER, Hilymax. Effectene or Altogen's cell line specific reagents.

Non Chemical Methods

- Electroporation is a popular method, although requiring an instrument and affecting the viability of many cell types, that also creates micro-sized holes transiently in the plasma membrane of cells under an electric discharge.

- Similarly, transfection applying sonic forces to cells, referred as Sono-poration.

- Optical transfection is a method where a tiny (~1 μm diameter) hole is transiently generated in the plasma membrane of a cell using a highly focused laser. This technique was first described in 1984 by Tsukakoshi et al., who used a frequency tripled Nd:YAG to generate stable and transient transfection of normal rat kidney cells. In this technique, one cell at a time is treated, making it particularly useful for single cell analysis.

- Gene electrotransfer is a technique that enables transfer of genetic material into prokaryotic or eukaryotic cells. It is based on a physical method named electroporation, where transient increase in the permeability of cell membrane is achieved when submitted to short and intense electric pulses.

- Impalefection is a method of introducing DNA bound to a surface of a nanofiber that is inserted into a cell.

Particle-based Methods

- A direct approach to transfection is the gene gun, where the DNA is coupled to a nanoparticle of an inert solid (commonly gold) which is then "shot" directly into the target cell's nucleus.

- Magnetofection, or Magnet assisted transfection is a transfection method, which uses magnetic force to deliver DNA into target cells. Nucleic acids are first associated with magnetic nanoparticles. Then, application of magnetic force drives the nucleic acid particle complexes towards and into the target cells, where the cargo is released.

* Impalefection is carried out by impaling cells by elongated nanostructures such as carbon nanofibers or silicon nanowires which have been functionalized with plasmid DNA.

Viral Methods

DNA can also be introduced into cells using viruses as a carrier. In such cases, the technique is called viral transduction, and the cells are said to be transduced. This can be done using insect cells.

Other (and Hybrid) Methods

Other methods of transfection include nucleofection, heat shock.

Stable and Transient Transfection

For most applications of transfection, it is sufficient if the transfected genetic material is only transiently expressed. Since the DNA introduced in the transfection process is usually not integrated into the nuclear genome, the foreign DNA will be diluted through mitosis or degraded.

If it is desired that the transfected gene actually remains in the genome of the cell and its daughter cells, a stable transfection must occur. To accomplish this, a marker gene is co-transfected, which gives the cell some selectable advantage, such as resistance towards a certain toxin. Some (very few) of the transfected cells will, by chance, have integrated the foreign genetic material into their genome.

If the toxin is then added to the cell culture, only those few cells with the marker gene integrated into their genomes will be able to proliferate, while other cells will die. After applying this selective stress (selection pressure) for some time, only the cells with a stable transfection remain and can be cultivated further. A common agent for selecting stable transfection is Geneticin, also known as G418, which is a toxin that can be neutralized by the product of the neomycin resistant gene.

RNA Transfection

RNA transfection is the process of deliberately introducing RNA into a living cell. RNA can be purified from cells after lysis or synthesized from free nucleotides either chemically, or enzymatically using an RNA polymerase to transcribe a DNA template. As with

DNA, RNA can be delivered to cells by a variety of means including microinjection, electroporation, and lipid-mediated transfection. If the RNA encodes a protein, transfected cells may translate the RNA into the encoded protein. If the RNA is a regulatory RNA (such as a miRNA), the RNA may cause other changes in the cell (such as RNAi-mediated knockdown).

Terminology

RNA molecules shorter than about 25nt largely evade detection by the innate immune system, which is triggered by longer RNA molecules. Most cells of the body express proteins of the innate immune system, and upon exposure to exogenous long RNA molecules these proteins initiate signalling cascades that result in inflammation. This inflammation hypersensitizes the exposed cell and nearby cells to subsequent exposure. As a result, while a cell can be repeatedly transfected with short RNA with few non-specific effects, repeatedly transfecting cells with even a small amount of long RNA can cause cell death unless measures are taken to suppress or evade the innate immune system.

Short-RNA Transfection

Short-RNA transfection is routinely used in biological research to knock down the expression of a protein of interest (using siRNA) or to express or block the activity of a miRNA (using short RNA that acts independently of the cell's RNAi machinery, and therefore is not referred to as siRNA). While DNA-based vectors (viruses, plasmids) that encode a short RNA molecule can also be used, short-RNA transfection does not risk modification of the cell's DNA, a characteristic that has led to the development of short RNA as a new class of macromolecular drugs.

Long-RNA Transfection

Long-RNA transfection is the process of deliberately introducing RNA molecules longer than about 25nt into living cells. A distinction is made between short- and long-RNA transfection because exogenous long RNA molecules elicit an innate immune response in cells that can cause a variety of nonspecific effects including translation block, cell-cycle arrest, and apoptosis.

Endogenous vs. Exogenous Long RNA

The innate immune system has evolved to protect against infection by detecting pathogen-associated molecular patterns (PAMPs), and triggering a complex set of responses collectively known as "inflammation". Many cells express specific pattern recognition receptors (PRRs) for exogenous RNA including toll-like receptor 3,7,8 (TLR3, TLR7, TLR8), the RNA helicase RIG1 (RARRES3), protein kinase R (PKR, a.k.a. EIF2AK2), members of the oligoadenylate synthetase family of proteins (OAS1, OAS2, OAS3), and others. All of these proteins can specifically bind to exogenous RNA molecules and trigger an immune response.

The specific chemical, structural or other characteristics of long RNA molecules that are required for recognition by PRRs remain largely unknown despite intense study. At any given time, a typical mammalian cell may contain several hundred thousand mRNA and other, regulatory long RNA molecules.

How cells distinguish exogenous long RNA from the large amount of endogenous long RNA is an important open question in cell biology. Several reports suggest that phosphorylation of the 5'-end of a long RNA molecule can influence its immunogenicity, and specifically that 5'-triphosphate RNA, which can be produced during viral infection, is more immunogenic than 5'-diphosphate RNA, 5'-monophosphate RNA or RNA containing no 5' phosphate. However, in vitro-transcribed (ivT) long RNA containing a 7-methylguanosine cap (present in eukaryotic mRNA) is also highly immunogenic despite having no 5' phosphate, suggesting that characteristics other than 5'-phosphorylation can influence the immunogenicity of an RNA molecule.

Eukaryotic mRNA contains chemically modified nucleotides such as N6-methyladenosine, 5-methylcytidine, and 2'-O-methylated nucleotides. Although only a very small number of these modified nucleotides are present in a typical mRNA molecule, they may help prevent mRNA from activating the innate immune system by disrupting secondary structure that would resemble double-stranded RNA (dsRNA), a type of RNA thought to be present in cells only during viral infection. The immunogenicity of long RNA has been used to study both innate and adaptive immunity.

Repeated Long-RNA Transfection

Inhibiting only three proteins, interferon-β, STAT2, and EIF2AK2 is sufficient to rescue human fibroblasts from the cell death caused by frequent transfection with long, protein-encoding RNA. Inhibiting interferon signalling disrupts the positive-feedback loop that normally hypersensitizes cells exposed to exogenous long RNA. Researchers have recently used this technique to express reprogramming proteins in primary human fibroblasts.

Transformation (Genetics)

In molecular biology transformation is the genetic alteration of a cell resulting from the direct uptake, incorporation and expression of exogenous genetic material (exogenous DNA) from its surrounding and taken up through the cell membrane(s). Transformation occurs most commonly in bacteria and in some species occurs naturally. Transformation can also be effected by artificial means.

Bacteria that are capable of being transformed, whether naturally or artificially, are called competent. Transformation is one of three processes by which exogenous genetic material may be introduced into bacterial cell, the other two being conjugation (transfer of genetic material between two bacterial cells in direct contact), and transduction (injection of foreign DNA by a bacteriophage into the host). Transformation may also used to describe the insertion of new genetic material into nonbacterial cells including animal and plant cells, however, because transformation has a special meaning in relation to animal cells indicating progression to a cancerous state, the term should be avoided for animal cells when describing introduction of exogenous genetic material. Introduction of foreign DNA into eukaryotic cells is usually called "transfection".

History

Transformation was first demonstrated in 1928 by British bacteriologist Frederick Griffith. Griffith discovered that a harmless strain of *Streptococcus pneumoniae* could be made virulent after being exposed to heat-killed virulent strains. Griffith hypothesized that some "transforming factor" from the heat-killed strain was responsible for making the harmless strain virulent. In 1944 this "transforming factor" was identified as being genetic by Oswald Avery, Colin MacLeod, and Maclyn McCarty. They isolated DNA from a virulent strain of *S. pneumoniae* and using just this DNA were able to make a harmless strain virulent. They called this uptake and incorporation of DNA by bacteria "transformation."

The results of Avery et al.'s experiments were at first sceptically received by the scientific community and it was not until the development of genetic markers and the discovery of other methods of genetic transfer (conjugation in 1947 and transduction in 1953) by

Joshua Lederberg that Avery's experiments were accepted. Transformation did not become routine procedure in laboratories until 1972 when Stanley Cohen, Annie Chang and Leslie Hsu successfully transformed *Escherichia coli* by treating the bacteria with calcium chloride. This created an efficient and convenient procedure for transforming bacteria and opened the way for molecular cloning in biotechnology and research.

Transformation using electroporation was developed in the late 1980s thus increasing the efficiency and number of bacterial strains that could be transformed. Transformation of animal and plant cells was also investigated with the first transgenic mouse being created by injecting a gene for a rat growth hormone into a mouse embryo in 1982. In 1907 a bacterium that caused plant tumours, *Agrobacterium tumefaciens*, was discovered and in the early 1970s the tumour inducing agent was found to be a DNA plasmid called the Ti plasmid. By removing the genes in the plasmid that caused the cancer and adding in novel genes researchers were able to infect plants with *A. tumefaciens* and let the bacteria insert their chosen DNA into the genomes of the plants. Not all plant cells are susceptible to infection by *A. tumefaciens* so other methods were developed including electroporation and micro-injection. Particle bombardment was made possible with the invention of the Biolistic Particle Delivery System (gene gun) by John Sanford in 1990.

Mechanisms

Bacteria

Bacteria transformation may be referred to as a stable genetic change brought about by the uptake of naked DNA (DNA without associated cells or proteins) and competence refers to the state of being able to take up exogenous DNA from the environment. Two forms of competence exist: natural and artificial.

Natural Competence

About 1% of bacterial species are capable of naturally taking up DNA under laboratory conditions; many more are able to take it up in their natural environments. Such bacteria carry sets of genes that provide the protein machinery to bring DNA across the cell membrane(s).

DNA material can be transferred between different strains of bacteria, in a process called horizontal gene transfer.

Artificial Competence

Artificial competence is induced by laboratory procedures and involves making the cell passively permeable to DNA by exposing it to conditions that do not normally occur in nature.

Calcium chloride transformation is a method of promoting competence. Chilling cells in the presence of divalent cations such as Ca (in $CaCl_2$) prepares the cell membrane to become permeable to plasmid DNA. The cells are incubated on ice with the DNA and then briefly heat shocked (e.g., 42°C for 30–120 seconds) thus allowing the DNA to enter the cells. This method works very well for circular plasmid DNA. An excellent preparation of competent cells will give ~10^8 colonies per microgram of plasmid. A poor preparation will be about $10^4/\mu g$ or less. Good, non-commercial preparations should give 10^5 to 10^6 transformants per microgram of plasmid. The method, however, usually does not work well for linear DNA, such as fragments of chromosomal DNA, probably because the cell's native exonuclease enzymes rapidly degrade linear DNA. Interestingly, cells that are naturally competent are usually transformed more efficiently with linear DNA than with plasmid DNA.

Electroporation is another method of promoting competence. In the method the cells are briefly shocked with an electric field of 10-20 kV/cm that creates holes in the cell membrane through which the plasmid DNA enters. This method is amenable to the uptake of large plasmid DNA. After the electric shock the holes are rapidly closed by the cell's membrane-repair mechanisms.

The efficiency with which a competent culture can take up exogenous DNA and express its genes is known as Transformation efficiency.

Plasmid Transformation

In order to be stably maintained in the cell a plasmid DNA molecule must contain an origin of replication, which allows it to be replicated in the cell independent of the replication of the cell's own chromosome. Because transformation usually produces a mixture of relatively few transformed cells and an abundance of non-transformed cells a method is needed to identify the cells that have acquired the plasmid. The method usually consists of using a plasmid that contains a gene that gives the bacterial cells resistance to an antibiotic that they are naturally sensitive to. The mixture of cells are then plated on media that contains the antibiotic thus only the transformed cells are able to grow. Cells that did not take up the plasmid are killed in the media. Another selection method called blue-white screen uses a plasmid that contains an antibiotic resistance gene and the *lacZ* gene. The *lacZ* gene codes for the *lacZ*-α subunit of the enzyme β-galactosidase, a homo-tetramer with each monomer composed of one *lacZ*-α subunit and one *lacZ*-ω subunit. The method also requires an *E. coli* strain that possesses in its genome the code for only the *lacZ*-ω subunit and not the *lacZ*-α subunit. One of the first steps in any transformation is the production of a recombined plasmid obtained by the successful ligation of the gene of interest into its corresponding vector, which in this method results in the disruption of *lacZ* because the gene of interest is inserted within the *lacZ* code. A cell that takes up a recombined plasmid would thus not be able to express the *lacZ*-α subunit and would, in turn, not be able to produce a functional β-galactosidase.

Conversely, a cell that has taken up non-recombined plasmid (perhaps one formed by the ligation of the vector's own two ends) will express the *lacZ*-α subunit and thus produce a functional β-galactosidase. A cell that does not take up any plasmid is not conferred with antibiotic resistance and will die upon plating. Consequently, the blue-white screen method allows for the ready detection of not just transformed cells, but, most importantly, cells that have been transformed by a successfully recombined plasmid. Selection occurs as a result of the action of β-galactosidase on its substrate X-gal, which is included in the media along with the appropriate antibiotic. X-gal is a colorless, modified galactose sugar whose hydrolysis by β-galactosidase produces galactose and the pre-chromophore 5-bromo-4-chloro-3-hydroxyindole. The latter is subsequently oxidized to 5,5'-dibromo-4,4'-dichloro-indigo, an insoluble, blue product that is readily seen by the naked eye. Colonies of cells that have been transformed by a successfully recombined plasmid will thus appear white whereas those that have been transformed by non-recombined plasmid will appear blue.

Plants

A number of mechanisms are available to transfer DNA into plant cells:

- *Agrobacterium* mediated transformation is the easiest and most simple plant transformation. Plant tissue (often leaves) are cut into small pieces, e.g. 10x10mm, and soaked for 10 minutes in a fluid containing suspended *Agrobacterium*. Some cells along the cut will be transformed by the bacterium, that inserts its DNA into the cell. Placed on selectable rooting and shooting media, the plants will regrow. Some plants species can be transformed just by dipping the flowers into suspension of *Agrobacterium* and then planting the seeds in a selective medium. Unfortunately, many plants are not transformable by this method.

- Particle bombardment: Particles of gold or tungsten are coated with DNA and then shot into young plant cells or plant embryos. Some genetic material will stay in the cells and transform them. This method also allows transformation of plant plastids. The transformation efficiency is lower than in agribacterial mediated transformation, but most plants can be transformed with this method.

- Electroporation: make transient holes in cell membranes using electric shock; this allows DNA to enter as described above for Bacteria.

- Viral transformation (transduction): Package the desired genetic material into a suitable plant virus and allow this modified virus to infect the plant. If the genetic material is DNA, it can recombine with the chromosomes to produce transformant cells. However genomes of most plant viruses consist of single stranded RNA which replicates in the cytoplasm of infected cell. For such genomes this method is a form of transfection and not a real transformation, since the inserted genes never reach the nucleus of the cell and do not integrate into the host genome. The progeny of the infected plants is virus free and also free of the inserted gene.

Animals

Introduction of DNA into animal cells is usually called transfection, and is discussed in the corresponding article.

Established Human Cell Lines

Cell lines that originate with humans have been somewhat controversial in bioethics, as they may outlive their parent organism and later be used in the discovery of lucrative medical treatments. In the pioneering decision in this area, the Supreme Court of California held in *Moore v. Regents of the University of California* that human patients have no property rights in cell lines derived from organs removed with their consent.

Generation of Hybridomas

It is possible to fuse normal cells with an immortalised cell line. This method is used to produce monoclonal antibodies. In brief, lymphocytes isolated from the spleen (or possibly blood) of an immunised animal are combined with an immortal myeloma cell line (B cell lineage) to produce a hybridoma which has the antibody specificity of the primary lymphoctye and the immortality of the myeloma. Selective growth medium (HA or HAT) is used to select against unfused myeloma cells; primary lymphoctyes die quickly in culture and only the fused cells survive. These are screened for production of the required antibody, generally in pools to start with and then after single cloning.

Applications of Cell Culture

Biological products produced by recombinant DNA (rDNA) technology in animal cell cultures include enzymes, synthetic hormones, immunobiologicals (monoclonal antibodies, interleukins, lymphokines), and anticancer agents. Although many simpler proteins can be produced using rDNA in bacterial cultures, more complex proteins that are glycosylated (carbohydrate-modified) currently must be made in animal cells. An important example of such a complex protein is the hormone erythropoietin. The cost of growing mammalian cell cultures is high, so research is underway to produce such complex proteins in insect cells or in higher plants, use of single embryonic cell and somatic embryos as a source for direct gene transfer via practicle bombardment, transit gene expression and confocal microscopy observation is one of its applications. It also offers to confirm single cell origin of somatic embryos and the asymmetry of the first cell division, which starts the process.

Tissue Culture and Engineering

Cell culture is a fundamental component of tissue culture and tissue engineering, as it establishes the basics of growing and maintaining cells *ex vivo*. The major application of human cell culture is in stem cell industry where mesenchymal stem cells can be cultured and cryopreserved for future use.

Vaccines

Vaccines for polio, measles, mumps, rubella, and chickenpox are currently made in cell cultures. Due to the H5N1 pandemic threat, research into using cell culture for influenza vaccines is being funded by the United States government. Novel ideas in the field include recombinant DNA-based vaccines, such as one made using human adenovirus (a common cold virus) as a vector,, such as adjuvants.

Culture of Non-mammalian Cells

Plant Cell Culture Methods: Plant cell cultures are typically grown as cell suspension cultures in liquid medium or as callus cultures on solid medium. The culturing of undifferentiated plant cells and calli requires the proper balance of the plant growth hormones auxin and cytokinin.

Microbiological Culture

A microbiological culture, or microbial culture, is a method of multiplying microbial organisms by letting them reproduce in predetermined culture media under controlled laboratory conditions. Microbial cultures are used to determine the type of organism, its abundance in the sample being tested, or both. It is one of the primary diagnostic methods of microbiology and used as a tool to determine the cause of infectious disease by letting the agent multiply in a predetermined medium. For example, a throat culture is taken by scraping the lining of tissue in the back of the throat and blotting the sample into a medium to be able to screen for harmful microorganisms, such as *Streptococcus pyogenes*, the causative agent of strep throat. Furthermore, the term culture is more generally used informally to refer to "selectively growing" a specific kind of microorganism in the lab.

Microbial cultures are foundational and basic diagnostic methods used extensively as a research tool in molecular biology. It is often essential to isolate a pure culture of microorganisms. A pure (or *axenic*) culture is a population of cells or multicellular organisms growing in the absence of other species or types. A pure culture may originate from a single cell or single organism, in which case the cells are genetic clones of one another. For the purpose of gelling the microbial culture, the medium of agarose gel (agar) is used. Agar is a gelatinous substance derived from seaweed. A cheap substitute for agar is guar gum, which can be used for the isolation and maintenance of thermophiles.

Bacterial Culture

Microbiological cultures use petri dishes of differing sizes that have a thin layer of agar-based growth medium in them. Once the growth medium in the petri dish is inoculated with the desired bacteria, the plates are incubated in an incubator (usually set at 37 degrees

Celsius for cultures from humans or animals, or lower for environmental cultures). Another method of bacterial culture is liquid culture, in which the desired bacteria are suspended in liquid broth, a nutrient medium. These are ideal for preparation of an antimicrobial assay. The experimenter would inoculate liquid broth with bacteria and let it grow overnight in a shaker for uniform growth, then take aliquots of the sample to test for the antimicrobial activity of a specific drug or protein (antimicrobial peptides).

Virus and Phage Culture

Virus or phage cultures require host cells in which the virus or phage multiply. For bacteriophages, cultures are grown by infecting bacterial cells. The phage can then be isolated from the resulting plaques in a lawn of bacteria on a plate. Virus cultures are obtained from their appropriate eukaryotic host cells.

Eukaryotic Cell Culture

Isolation of Pure Cultures: For single-celled eukaryotes, such as yeast, the isolation of pure cultures uses the same techniques as for bacterial cultures. Pure cultures of multicellular organisms are often more easily isolated by simply picking out a single individual to initiate a culture. This is a useful technique for pure culture of fungi, multicellular algae, and small metazoa, for example.

Developing pure culture techniques is crucial to the observation of the specimen in question. The most common method to isolate individual cells and produce a pure culture is to prepare a streak plate. The streak plate method is a way to physically separate the microbial population, and is done by spreading the inoculate back and forth with an inoculating loop over the solid agar plate. Upon incubation, colonies will arise and, hopefully, single cells will have been isolated from the biomass.

Viral Culture Methods

The culture of viruses requires the culture of cells of mammalian, plant, fungal or bacterial origin as hosts for the growth and replication of the virus. Whole wild type viruses, recombinant viruses or viral products may be generated in cell types other than their natural hosts under the right conditions. Depending on the species of the virus, infection and viral replication may result in host cell lysis and formation of a viral plaque.

Organ Culture

Organ culture is a development from tissue culture methods of research, the organ culture is able to accurately model functions of an organ in various states and conditions by the use of the actual *in vitro* organ itself.

Parts of an organ or a whole organ can be cultured in vitro. The main objective is to maintain the architecture of the tissue and direct it towards normal development. In this technique, it is essential that the tissue is never be disrupted or damaged. It thus requires careful handling. The media used for a growing organ culture are generally the same as those used for tissue culture. The techniques for organ culture can be classified into (i) those employing a solid medium and (ii) those employing liquid medium.

Methodology

Embryonic organ culture is an easier alternative to normal organ culture derived from adult animals. The following are three techniques employed for embryonic organ culture.

Plasma Clot Method

The following are general steps in organ culture on plasma clots.

1. Prepare a plasma clot by mixing 15 drops of plasma with five drops of embryo extract in a watch glass.
2. Place a watch glass on a pad of cotton wool in a petri dish; cotton wool is kept moist to prevent excessive evaporation from the dish.
3. Place a small, carefully dissected piece of tissue on top of the plasma clots in watch glass.

The technique has now been modified, and a raft of lens paper or rayon net is used on which the tissue is placed. Transfer of the tissue can then be achieved by raft easily. Excessive fluid is removed and the net with the tissue placed again on the fresh pool of medium.Parts of an organ or a whole organ can be cultured in vitro. The main objective is to maintain the architecture of the tissue and direct it towards normal development. In this technique, it is essential that the tissue is never be disrupted or damaged.

It thus requires careful handling. The media used for a growing organ culture are generally the same as those used for tissue culture. The techniques for organ culture can be classified into (i) those employing a solid medium and (ii) those employing liquid medium.

Agar Gel Method

Media solidified with agar are also used for organ culture and these media consist of 7 parts 1% agar in BSS, 3 parts chick embryo extract and 3 parts of horse serum. Defined media with or without serum are also used with agar. The medium with agar provides the mechanical support for organ culture. It does not liquefy. Embryonic organs generally grow well on agar, but adult organ culture will not survive on this medium.

The culture of adult organs or parts from adult animals is more difficult due to their greater requirement of oxygen. A variety of adult organs (e.g. the liver) have been cultured using special media with special apparatus (Towell's II culture chamber). Since serum was found to be toxic, serum-free media were used, and the special apparatus permitted the use of 95% oxygen.

Raft Methods In this approach the explant is placed onto a raft of lens paper or rayon acetate, which is floated on serum in a watch glass. Rayon acetate rafts are made to float on the serum by treating their 4 corners with silicone. Similarly, floatability of lens paper is enhanced by treating it with silicone. On each raft, 4 or more explants are usually placed.

In a combination of raft and clot techniques, the explants are first placed on a suitable raft, which is then kept on a plasma clot. This modification makes media changes easy, and prevents the sinking of explants into liquefied plasma.

Grid Method

Initially devised by Trowell in 1954, the grid method utilizes 25 mm x 25 mm pieces of a suitable wire mesh or perforated stainless steel sheet whose. edges are bent to form 4 legs of about 4 mm height.

Skeletal tissues are generally placed directly on the grid but softer tissues like glands or skin are first placed on rafts, which are then kept on the grids. The grids themselves are placed in a culture chamber filled with fluid medium up to the grid; the chamber is supplied with a mixture of O_2 and CO_2 to meet the high O_2 requirements of adult mammalian organs. A modification of the original grid method is widely used to study the growth and differentiation of adult and embryonic tissues.

Limitations

- Results from organ cultures are often not comparable to those from whole animals studies, e.g. in studies on drug action, since the drug are metabolized in vivo but not in vitro.

Current Progress

In April 2006, scientists reported a successful trial of seven bladders grown in-vitro and given to humans.

Bioeconomics

Bioeconomics is the study of the dynamics of living resources using economic models. It is an attempt to apply the methods of environmental economics and ecological economics to empirical biology. Bioeconomics applies optimal control methods to mathematical models using environmental and ecological elements for resource protection issues relating to resource economics. Bioeconomics is the science determining the socioeconomic activity threshold for which a biological system can be effectively and efficiently utilised without destroying the conditions for its regeneration and therefore its sustainability.

Bioeconomics is closely related to the early development of theories in fisheries economics, initially in the mid 1950s by Canadian economists Scott Gordon (in 1954) and Anthony Scott (1955). Their ideas used recent achievements in biological fisheries modelling, primarily the works by Schaefer (1957) on establishing a formal relationship between fishing activities and biological growth through mathematical modelling confirmed by empirical studies, and also relates itself to ecology and the environment and resource protection. These ideas developed out of the multidisciplinary fisheries science environment in Canada at the time. Fisheries science and modelling developed rapidly during a productive and innovative period, particularly among Canadian fisheries researchers of various disciplines. Population modelling and fishing mortality were introduced to economists, and new interdisciplinary modelling tools became available for the economists, which made it possible to evaluate biological and economic impacts of different fishing activities and fisheries management decisions.

Renewable Resources

Fisheries: At least one researcher has attempted to perform Hubbert linearization (Hubbert curve) on the whaling industry, as well as charting the transparently dependent price of caviar on sturgeon depletion. Another example is the cod of the North Sea. The comparison of the cases of fisheries and of mineral extraction tells us that the human pressure on the environment is causing a wide range of resources to go through a depletion cycle which follows a Hubbert curve.

Biopolitics

The term "biopolitics" or "biopolitical" can refer to several different yet compatible concepts.

Definitions

1. In the work of Michel Foucault, the style of government that regulates populations through biopower (the application and impact of political power on all aspects of human life).

2. In the works of Michael Hardt and Antonio Negri, anti-capitalist insurrection using life and the body as weapons; examples include flight from power and, 'in its most tragic and revolting form', suicide terrorism. Conceptualised as the opposite of biopower, which is seen as the practice of sovereignty in biopolitical conditions.

3. The political application of bioethics.

4. A political spectrum that reflects positions towards the sociopolitical consequences of the biotech revolution.

5. Political advocacy in support of, or in opposition to, some applications of biotechnology.

6. Public policies regarding some applications of biotechnology.

7. Political advocacy concerned with the welfare of all forms of life and how they are moved by one another.

Politics and the Life Sciences

As a field of the academic discipline of political science, biopolitics is also known as "politics and the life sciences". The Association for Politics and the Life Sciences was formed in 1981 and exists to study the field of biopolitics as a subfield of political science. APLS owns and publishes an academic peer-reviewed journal called *Politics and the Life Sciences* (PLS). The journal is edited in the United States at the University of Maryland, College Park's School of Public Policy, in Maryland. The Department of Political Science at Northern Illinois University offers undergraduate and graduate courses in the field of politics and the life sciences. It is the only political science department in the U.S. to offer politics and the life sciences as a graduate field of study.

Genetic Engineering

Genetic engineering, also called genetic modification, is the direct human manipulation of an organism's genetic material in a way that does not occur under natural conditions. It involves the use of recombinant DNA techniques, but does not include traditional animal and plant breeding or mutagenesis. Any organism that is generated using these techniques is considered to be a genetically modified organism. The first organisms genetically engineered were bacteria in 1973 and then mice in 1974. Insulin producing bacteria were commercialized in 1982 and genetically modified food has been sold since 1994.

The most common form of genetic engineering involves the insertion of new genetic material at an unspecified location in the host genome. This is accomplished by isolating and copying the genetic material of interest, generating a construct containing all the genetic elements for correct expression, and then inserting this construct into the host organism. Other forms of genetic engineering include gene targeting and knocking out specific genes via engineered nucleases such as zinc finger nucleases or engineered homing endonucleases. Genetic engineering techniques have been applied in numerous fields including research, biotechnology, and medicine. Medicines such as insulin and human growth hormone are now produced in bacteria, experimental mice such as the oncomouse and the knockout mouse are being used for research purposes and insect resistant and/or herbicide tolerant crops have been commercialized. Genetically engineered plants and animals capable of producing biotechnology drugs more cheaply than current methods (called pharming) are also being developed and in 2009 the FDA approved the sale of the pharmaceutical protein antithrombin produced in the milk of genetically engineered goats.

Definition

Genetic engineering alters the genetic makeup of an organism using techniques that introduce heritable material prepared outside the organism either directly into the host or into a cell that is then fused or hybridized with the host. This involves using recombinant nucleic acid (DNA or RNA) techniques to form new combinations of heritable genetic material followed by the incorporation of that material either indirectly through a vector system or directly through micro-injection, macro-injection and micro-encapsulation techniques. Genetic engineering does not include traditional animal and plant breeding, in vitro fertilisation, induction of polyploidy, mutagenesis and cell fusion techniques that do not use recombinant nucleic acids or a genetically modified organism in the process. Cloning and stem cell research, although not considered genetic engineering, are closely related and genetic engineering can be used within them. Synthetic biology is an emerging discipline that takes genetic engineering a step further by introducing artificially synthesized genetic material from raw materials into an organism. If genetic material from another species is added to the host, the resulting organism is called transgenic. If genetic material from the same species or a species that can naturally breed with the host is used the resulting organism is called cisgenic. Genetic engineering can also be used to remove genetic material from the target organism, creating a knock out organism. In Europe genetic modification is synonymous with genetic engineering while within the United States of America it can also refer to conventional breeding methods.

Humans have altered the genomes of species for thousands of years through artificial selection and more recently mutagenesis. Genetic engineering as the direct manipulation of DNA by humans outside breeding and mutations has only existed since the 1970s. The term "genetic engineering" was first coined by Jack Williamson in his science fiction novel *Dragon s Island*, published in 1951, one year before DNA's role in heredity was confirmed by Alfred Hershey and Martha Chase, and two years before James Watson and Francis Crick showed that the DNA molecule has a double-helix structure.

In 1972 Paul Berg created the first recombinant DNA molecules by combined DNA from the monkey virus SV40 with that of the lambda virus. In 1973 Herbert Boyer and Stanley Cohen created the first transgenic organism by inserting antibiotic resistance genes into the plasmid of an *E. coli* bacterium. A year later Rudolf Jaenisch created a transgenic mouse by introducing foreign DNA into its embryo, making it the world's first transgenic animal. In 1976 Genentech, the first genetic engineering company was founded by Herbert Boyer and Robert Swanson and a year later and the company produced a human protein (somatostatin) in *E.coli*. Genentech announced the production of genetically engineered human insulin in 1978. In 1980, the U.S. Supreme Court in the Diamond v. Chakrabarty case ruled that genetically altered life could be patented. The insulin produced by bacteria, branded humulin, was approved for release by the Food and Drug Administration in 1982.

The first field trials of genetically engineered plants occurred in France and the USA in 1986, tobacco plants were engineered to be resistant to herbicides. The People's Republic of China was the first country to commercialize transgenic plants, introducing a virus-resistant tobacco in 1992. In 1994 Calgene attained approval to commercially release the Flavr Savr tomato, a tomato engineered to have a longer shelf life. In 1994, the European Union approved tobacco engineered to be resistant to the herbicide bromoxynil, making it the first genetically engineered crop commercialized in Europe. In 1995, Bt Potato was approved safe by the Environmental Protection Agency, making it the first pesticide producing crop to be approved in the USA. In 2009 11 transgenic crops were grown commercially in 25 countries, the largest of which by area grown were the USA, Brazil, Argentina, India, Canada, China, Paraguay and South Africa.

In 2010, scientists at the J. Craig Venter Institute, announced that they had created the first synthetic bacterial genome, and added it to a cell containing no DNA. The resulting bacterium, named Synthia, was the world's first synthetic life form.

Process

Isolating the Gene

First, the gene to be inserted into the genetically modified organism must be chosen and isolated. Presently, most genes transferred into plants provide protection against insects or tolerance to herbicides. In animals the majority of genes used are growth hormone genes. Once chosen the genes must be isolated. This typically involves multiplying the gene using polymerase chain reaction (PCR). If the chosen gene or the donor organism's genome has been well studied it may be present in a genetic library. If the DNA sequence is known, but no copies of the gene are available, it can be artificially synthesized. Once isolated, the gene is inserted into a bacterial plasmid.

Constructs

The gene to be inserted into the genetically modified organism must be combined with other genetic elements in order for it to work properly. The gene can also be modified at this stage for better expression or effectiveness. As well as the gene to be inserted most constructs contain a promoter and terminator region as well as a selectable marker gene. The promoter region initiates transcription of the gene and can be used to control the location and level of gene expression, while the terminator region ends transcription. The selectable marker, which in most cases confers antibiotic resistance to the organism it is expressed in, is needed to determine which cells are transformed with the new gene. The constructs are made using recombinant DNA techniques, such as restriction digests, ligations and molecular cloning.

Gene Targeting

Gene targeting (also, replacement strategy based on homologous recombination) is a genetic technique that uses homologous recombination to change an endogenous gene. The method can be used to delete a gene, remove exons, add a gene, and introduce point mutations. Gene targeting can be permanent or conditional. Conditions can be a specific time during development/life of the organism or limitation to a specific tissue, for example. Gene targeting requires the creation of a specific vector for each gene of interest. However, it can be used for any gene, regardless of transcriptional activity or gene size.

Methods

Gene targeting methods are established for several model organisms and may vary depending on the species used. In general, a targeting construct made out of DNA is generated in bacteria. It typically contains part of the gene to be targeted, a reporter gene, and a (dominant) selectable marker. To target genes in mice, this construct is then inserted into mouse embryonic stem cells in culture. After cells with the correct insertion have been selected, they can be used to contribute to a mouse's tissue via embryo injection. Finally, chimeric mice where the modified cells made up the reproductive organs are selected for via breeding. After this step the entire body of the mouse is based on the previously selected embryonic stem cell. To target genes in moss, this construct is incubated together with freshly isolated protoplasts and with Polyethylene glycol. As mosses are haploid organisms, regenerating moss filaments (protonema) can directly be screened for gene targeting, either by treatment with antibiotics or with PCR. Unique among plants, this procedure for reverse genetics is as efficient as in yeast. Using modified procedures, gene targeting has also been successfully applied to cattle, sheep, swine, and many fungi.

Comparison with Gene Trapping

Gene trapping is based on random insertion of a cassette while gene targeting targets a specific gene. Cassettes can be used for many different things while the flanking homology regions of gene targeting cassettes need to be adapted for each gene. This makes gene trapping more easily amenable for large scale projects than targeting. On the other hand, gene targeting can be used for genes with low transcriptions that would go undetected in a trap screen. Also, the probability of trapping increases with intron size. For gene targeting these compact genes are just as easily altered.

Applications

Gene targeting has been widely used to study human genetic diseases by removing "knock-out", or adding "knock-in", specific mutations of interest to a variety of models. Previously used to engineer rat cell models, advances in gene targeting technologies are enabling the creation of a new wave of isogenic human disease models. These models are the most accurate in-vitro models available to researchers to date, and are facilitating the development of new personalised drugs and diagnostics, particularly in the field of cancer.

2007 Nobel Prize

Mario R. Capecchi, Martin J. Evans and Oliver Smithies were declared laureates of the 2007 Nobel Prize in Physiology or Medicine for their work on "principles for introducing specific gene modifications in mice by the use of embryonic stem cells", or gene targeting.

Selection

Not all the organism's cells will be transformed with the new genetic material; in most cases a selectable marker is used to differentiate transformed from untransformed cells. If a cell has been successfully transformed with the DNA it will also contain the marker gene. By growing the cells in the presence of an antibiotic or chemical that selects or marks the cells expressing that gene it is possible to separate the transgenic events from the non-transgenic. Another method of screening involves using a DNA probe that will only stick to the inserted gene. A number of strategies have been developed that can remove the selectable marker from the mature transgenic plant.

Regeneration

As often only a single cell is transformed with genetic material the organism must be regrown from that single cell. As bacteria consist of a single cell and reproduce clonally regeneration is not necessary. In plants this is accomplished through the use of tissue culture. Each plant species has different requirements for successful regeneration through tissue culture.

If successful an adult plant is produced that contains the transgene in every cell. In animals it is necessary to ensure that the inserted

DNA is present in the embryonic stem cells. When the offspring is produced they can be screened for the presence of the gene. All offspring from the first generation will be heterozygous for the inserted gene and must be mated together to produce a homozygous animal.

Confirmation

Further tests using PCR, Southern Blots and Bioassays are needed to confirm that the gene is expressed and functions correctly. The organism's offspring are also tested to ensure that the trait can be inherited and that it follows a Mendelian inheritance pattern.

Applications

Genetic engineering has applications in medicine, research, industry and agriculture and can be used on a wide range of plants, animals and micro organism.

Medicine

In medicine genetic engineering has been used to mass produce insulin, human growth hormones, follistim (for treating infertility), human albumin, monoclonal antibodies, antihemophilic factors, vaccines and many other drugs. Vaccination generally involves injecting weak live, killed or inactivated forms of viruses or their toxins into the person being immunized. Genetically engineered viruses are being developed that can still confer immunity, but lack the infectious sequences. Mouse hybridomas, cells fused together to create monoclonal antibodies, have been humanised through genetic engineering to create human monoclonal antibodies.

Genetic engineering is used to create animal models of human diseases. Genetically modified mice are the most common genetically engineered animal model. They have been used to study and model cancer (the oncomouse), obesity, heart disease, diabetes, arthritis, substance abuse, anxiety, aging and Parkinson disease. Potential cures can be tested against these mouse models. Also genetically modified pigs have been bred with the aim of increasing the success of pig to human organ transplantation.

Gene therapy is the genetic engineering of humans by replacing defective human genes with functional copies. This can occur in somatic tissue or germline tissue. If the gene is inserted into the germline tissue it can be passed down to that person's descendants. Gene therapy has been used to treat patients suffering from immune deficiencies (notably Severe combined immunodeficiency) and trials have been carried out on other genetic disorders.

The success of gene therapy so far has been limited and a patient (Jesse Gelsinger) has died during a clinical trial testing a new treatment. There are also ethical concerns should the technology be used not just for treatment, but for enhancement, modification or alteration of a human beings' appearance, adaptability, intelligence, character or Behaviour. The distinction between cure and enhancement can also be difficult to establish. Transhumanists consider the enhancement of humans desirable.

Research

Genetic engineering is an important tool for natural scientists. Genes and other genetic information from a wide range of organisms are transformed into bacteria for storage and modification, creating genetically modified bacteria in the process. Bacteria are cheap, easy to grow, clonal, multiply quickly, relatively easy to transform and can be stored at -80°C almost indefinitely. Once a gene is isolated it can be stored inside the bacteria providing an unlimited supply for research.

Organisms are genetically engineered to discover the functions of certain genes. This could be the effect on the phenotype of the organism, where the gene is expressed or what other genes it interacts with. These experiments generally involve loss of function, gain of function, tracking and expression.

- Loss of function experiments, such as in a gene knockout experiment, in which an organism is engineered to lack the activity of one or more genes. A knockout experiment involves the creation and manipulation of a DNA construct *in vitro*, which, in a simple knockout, consists of a copy of the desired gene, which has been altered such that it is non-functional. Embryonic stem cells incorporate the altered gene, which replaces the already present functional copy. These stem cells are injected into blastocysts, which are implanted into surrogate mothers. This allows the experimenter to analyse the defects caused by this mutation and thereby determine the role of particular genes. It is used especially frequently in developmental biology. Another method, useful in organisms such as Drosophila (fruit fly), is to induce mutations in a large population and then screen the progeny for the desired mutation. A similar process can be used in both plants and prokaryotes.

- Gain of function experiments, the logical counterpart of knockouts. These are sometimes performed in conjunction with knockout experiments to more finely establish the function of the desired gene. The process is much the same as that in knockout engineering, except that the construct is designed to increase the function of the gene, usually by providing extra copies of the gene or inducing synthesis of the protein more frequently.

- Tracking experiments, which seek to gain information about the localization and interaction of the desired protein. One way to do this is to replace the wild-type gene with a 'fusion' gene, which is a juxtaposition of the wild-type gene with a reporting element such as green fluorescent protein (GFP) that will allow easy visualization of the products of the genetic modification. While this is a useful technique, the manipulation can destroy the function of the gene, creating secondary effects and possibly calling into question the results of the experiment. More sophisticated techniques are now in development that can track protein products without mitigating their function, such as the addition of small sequences that will serve as binding motifs to monoclonal antibodies.

- Expression studies aim to discover where and when specific proteins are produced. In these experiments, the DNA sequence before the DNA that codes for a protein, known as a gene's promoter, is reintroduced into an organism with the protein coding region replaced by a reporter gene such as GFP or an enzyme that catalyses the production of a dye. Thus the time and place where a particular protein is produced can be observed. Expression studies can be taken a step further by altering the promoter to find which pieces are crucial for the proper expression of the gene and are actually bound by transcription factor proteins; this process is known as promoter bashing.

Industrial

By engineering genes into bacterial plasmids it is possible to create a biological factory that can produce proteins and enzymes. Some genes do not work well in bacteria, so yeast, a eukaryote, can also be used. Bacteria and yeast factories have been used to produce medicines such as insulin, human growth hormone, and vaccines, supplements such as tryptophan, aid in the production of food (chymosin in cheese making) and fuels. Other applications involving genetically engineered bacteria being investigated involve making the bacteria perform tasks outside their natural cycle, such as cleaning up oil spills, carbon and other toxic waste.

Agriculture

One of the best-known and controversial applications of genetic engineering is the creation of genetically modified food. There are three generations of genetically modified crops. First generation crops have been commercialized and most provide protection from insects and/or resistance to herbicides. There are also fungal and virus resistant crops developed or in development. They have been developed to make the insect and weed management of crops easier and can indirectly increase crop yield.

The second generation of genetically modified crops being developed aim to directly improve yield by improving salt, cold or drought tolerance and to increase the nutritional value of the crops. The third generation consists of pharmaceutical crops, crops that contain edible vaccines and other drugs. Some agriculturally important animals have been genetically modified with growth hormones to increase their size while others have been engineered to express drugs and other proteins in their milk.

The genetic engineering of agricultural crops can increase the growth rates and resistance to different diseases caused by pathogens and parasites. This is beneficial as it can greatly increase the production of food sources with the usage of fewer resources that would be required to host the world's growing populations. These modified crops would also reduce the usage of chemicals, such as fertilizers and pesticides, and therefore decrease the severity and frequency of the damages produced by these chemical pollution.

Ethical and safety concerns have been raised around the use of genetically modified food. A major safety concern relates to the human health implications of eating genetically modified food, in particular whether toxic or allergic reactions could occur. Gene flow into related non-transgenic crops, off target effects on beneficial organisms and the impact on biodiversity are important environmental issues. Ethical concerns involve religious issues, corporate control of the food supply, intellectual property rights and the level of labelling needed on genetically modified products.

2

Comparative Genomics

Comparative genomics is the study of the relationship of genome structure and function across different biological species or strains. Comparative genomics is an attempt to take advantage of the information provided by the signatures of selection to understand the function and evolutionary processes that act on genomes. While it is still a young field, it holds great promise to yield insights into many aspects of the evolution of modern species. The sheer amount of information contained in modern genomes (3.2 gigabases in the case of humans) necessitates that the methods of comparative genomics are automated. Gene finding is an important application of comparative genomics, as is discovery of new, non-coding functional elements of the genome.

Comparative genomics exploits both similarities and differences in the proteins, RNA, and regulatory regions of different organisms to infer how selection has acted upon these elements. Those elements that are responsible for similarities between different species should be conserved through time (stabilizing selection), while those elements responsible for differences among species should be divergent (positive selection). Finally, those elements that are unimportant to the evolutionary success of the organism will be unconserved (selection is neutral). One of the important goals of the field is the identification of the mechanisms of eukaryotic genome evolution. It is however often complicated by the multiplicity of events that have taken place throughout the history of individual lineages, leaving only distorted and superimposed traces in the genome of each living organism. For this reason comparative genomics studies of small model organisms (for example yeast) are of great importance to advance our understanding of general mechanisms of evolution.

Having come a long way from its initial use of finding functional proteins, comparative genomics is now concentrating on finding regulatory regions and siRNA molecules. Recently, it has been discovered that distantly related species often share long conserved stretches of DNA that do not appear to code for any protein. One such ultra-conserved region, that was stable from chicken to chimp has undergone a sudden burst of change in the human lineage, and is found to be active in the developing brain of the human embryo.

Computational approaches to genome comparison have recently become a common research topic in computer science. A public collection of case studies and demonstrations is growing, ranging from whole genome comparisons to gene expression analysis. This has increased the introduction of different ideas, including concepts from systems and control, information theory, strings analysis and data mining. It is anticipated that computational approaches will become and remain a standard topic for research and teaching, while multiple courses will begin training students to be fluent in both topics.

Modelling Biological Systems

Modelling biological systems is a significant task of systems biology and mathematical biology. Computational systems biology aims to develop and use efficient algorithms, data structures, visualization and communication tools to orchestrate the integration of large quantities of biological data with the goal of computer modelling. It involves the use of computer simulations of biological systems, like cellular subsystems (such as the networks of metabolites and enzymes which comprise metabolism, signal transduction pathways and gene regulatory networks) to both analyse and visualize the complex connections of these cellular processes.

It is also directly associated with bioinformatics and computational biology. Artificial life or virtual evolution attempts to understand evolutionary processes via the computer simulation of simple (artificial) life forms.

Overview

It is understood that an unexpected emergent property of a complex system is a result of the interplay of the cause-and-effect among simpler, integrated parts. Biological systems manifest many important examples of emergent properties in the complex interplay of components. Traditional study of biological systems requires reductive methods in which quantities of data are gathered by category, such as concentration over time in response to a certain stimulus. Computers are critical to analysis and modelling of these data. The goal is to create accurate real-time models of a system's response to environmental and internal stimuli, such as a model of a cancer cell in order to find weaknesses in its signalling pathways, or modelling of ion channel mutations to see effects on cardiomyocytes and in turn, the function of a beating heart.

A monograph on this topic summarizes an extensive amount of published research in this area up to 1987, including subsections in the following areas: computer modelling in biology and medicine, arterial system models, neuron models, biochemical and oscillation networks, quantum automata, quantum computers in molecular biology and genetics, cancer modelling, neural nets, genetic networks, abstract relational biology, metabolic-replication systems, category theory applications in biology and medicine, automata theory, cellular automata, tessallation models and complete self-reproduction, chaotic systems in organisms, relational biology and organismic theories. This published report also includes 390 references to peer-reviewed articles by a large number of authors.

Standards

By far the most widely accepted standard format for storing and exchanging models in the field is the Systems Biology Markup Language (SBML) The SBML.org website includes a guide to many important software packages used in computational systems biology. Other markup languages with different emphases include BioPAX and CellML.

Cellular Model

Creating a cellular model has been a particularly challenging task of systems biology and mathematical biology. It involves developing efficient algorithms, data structures, visualization and communication tools to orchestrate the integration of large quantities of biological data with the goal of computer modelling. It is also directly associated with bioinformatics, computational biology and Artificial life.

It involves the use of computer simulations of the many cellular subsystems such as the networks of metabolites and enzymes which comprise metabolism, signal transduction pathways and gene regulatory networks to both analyse and visualize the complex connections of these cellular processes.

The complex network of biochemical reaction/transport processes and their spatial organization make the development of a predictive model of a living cell a grand challenge for the 21st century.

Overview

The eukaryotic cell cycle is very complex and is one of the most studied topics, since its misregulation leads to cancers. It is possibly a good example of a mathematical model as it deals with simple calculus but gives valid results. Two research groups have produced several models of the cell cycle simulating several organisms. They have recently produced a generic eukaryotic cell cycle model which can represent a particular eukaryote depending on the values of the parameters, demonstrating that the idiosyncrasies of the individual cell cycles are due to different protein concentrations and affinities, while the underlying mechanisms are conserved.

By means of a system of ordinary differential equations these models show the change in time (dynamical system) of the protein inside a single typical cell; this type of model is called a deterministic process (whereas a model describing a statistical distribution of protein concentrations in a population of cells is called a stochastic process).

To obtain these equations an iterative series of steps must be done: first the several models and observations are combined to form a consensus diagram and the appropriate kinetic laws are chosen to write the differential equations, such as rate kinetics for stoichiometric reactions, Michaelis-Menten kinetics for enzyme substrate reactions and Goldbeter–Koshland kinetics for ultrasensitive transcription factors, afterwards the parameters of the equations (rate constants, enzyme efficiency coefficients and Michealis constants) must be fitted to match observations; when they cannot be fitted the kinetic equation is revised and when that is not possible the wiring diagram is modified. The parameters are fitted and validated using observations of both wild type and mutants, such as protein half-life and cell size. In order to fit the parameters the differential equations need to be studied. This can be done either by simulation or by analysis.

In a simulation, given a starting vector (list of the values of the variables), the progression of the system is calculated by solving the equations at each time-frame in small increments.

In analysis, the proprieties of the equations are used to investigate the Behaviour of the system depending of the values of the parameters and variables. A system of differential equations can be represented as a vector field, where each vector described the change (in concentration of two or more protein) determining where and how fast the trajectory (simulation) is heading. Vector fields can have several special points: a stable point, called a sink, that attracts in all directions (forcing the concentrations to be at a certain value), an unstable point, either a source or a saddle point which repels (forcing the concentrations to change away from a certain value), and a limit cycle, a closed trajectory towards which several trajectories spiral towards (making the concentrations oscillate).

A better representation which can handle the large number of variables and parameters is called a bifurcation diagram (bifurcation theory): the presence of these special steady-state points at certain values of a parameter (e.g. mass) is represented by a point and once the parameter passes a certain value, a qualitative change occurs, called a bifurcation, in which the nature of the space changes, with profound consequences for the protein concentrations: the cell cycle has phases (partially corresponding to G1 and G2) in which mass, via a stable point, controls cyclin levels, and phases (S and M phases) in which the concentrations change independently, but once the phase has changed at a bifurcation event (cell cycle checkpoint), the system cannot go back to the previous levels since at the current mass the vector field is profoundly different and the mass cannot be reversed back through the bifurcation event, making a checkpoint irreversible. In particular the S and M checkpoints are regulated by means of special bifurcations called a Hopf bifurcation and an infinite period bifurcation.

Protein Structure Prediction

Protein structure prediction is the prediction of the three-dimensional structure of a protein from its amino acid sequence — that is, the prediction of its secondary, tertiary, and quaternary structure from its primary structure. Structure prediction is fundamentally different from the inverse problem of protein design. Protein structure prediction is one of the most important goals pursued by bioinformatics and theoretical chemistry; it is highly important in medicine (for example, in drug design) and biotechnology (for example, in the design of novel enzymes). Every two years, the performance of current methods is assessed in the CASP experiment (Critical Assessment of Techniques for Protein Structure Prediction).

Secondary Structure

Secondary structure prediction is a set of techniques in bioinformatics that aim to predict the local secondary structures of proteins and RNA sequences based only on knowledge of their primary structure-amino acid or nucleotide sequence, respectively. For proteins, a prediction consists of assigning regions of the amino acid sequence as likely alpha helices, beta strands (often noted as "extended" conformations), or turns. The success of a prediction is determined by comparing it to the results of the DSSP algorithm applied to the crystal structure of the protein; for nucleic acids, it may be determined from the hydrogen bonding pattern. Specialized algorithms have been developed for the detection of specific well-defined patterns such as transmembrane helices and coiled coils in proteins, or canonical microRNA structures in RNA.

The best modern methods of secondary structure prediction in proteins reach about 80% accuracy; this high accuracy allows the use of the predictions in fold recognition and ab initio protein structure prediction, classification of structural motifs, and refinement of sequence alignments. The accuracy of current protein secondary structure prediction methods is assessed in weekly benchmarks such as LiveBench and EVA.

Background

Early methods of secondary structure prediction, introduced in the 1960s and early 1970s, focused on identifying likely alpha helices and were based mainly on helix-coil transition models. Significantly more accurate predictions that included beta sheets were introduced in the 1970s and relied on statistical assessments based on probability parameters derived from known solved structures. These methods, applied to a single sequence, are typically at most about 60-65% accurate, and often underpredict beta sheets.

The evolutionary conservation of secondary structures can be exploited by simultaneously assessing many homologous sequences in a multiple sequence alignment, by calculating the net secondary structure propensity of an aligned column of amino acids. In concert with larger databases of known protein structures and modern machine learning methods such as neural nets and support vector machines, these methods can achieve up 80% overall accuracy in globular proteins.

The theoretical upper limit of accuracy is around 90%, partly due to idiosyncrasies in DSSP assignment near the ends of secondary structures, where local conformations vary under native conditions but may be forced to assume a single conformation in crystals due to packing constraints. Limitations are also imposed by secondary structure prediction's inability to account for tertiary structure; for example, a sequence predicted as a likely helix may still be able to adopt a beta-strand conformation if it is located within a beta-sheet region of the protein and its side chains pack well with their neighbours. Dramatic conformational changes related to the protein's function or environment can also alter local secondary structure.

Chou-Fasman Method

The Chou-Fasman method are an empirical technique for the prediction of secondary structures in proteins, originally developed in the 1970s. The method is based on analyses of the relative frequencies of each amino acid in alpha helices, beta sheets, and turns based on known protein structures solved with X-ray crystallography. From these frequencies a set of probability parameters were derived for the appearance of each amino acid in each secondary structure type, and these parameters are used to predict the probability that a given sequence of amino acids would form a helix, a beta strand, or a turn in a protein. The method is at most about 50-60% accurate in identifying correct secondary structures, which is significantly less accurate than the GOR method or modern machine learning-based techniques.

Amino Acid Propensities

The original Chou-Fasman parameters found some strong tendencies among individual amino acids to prefer one type of secondary structure over others. Alanine, glutamate, leucine, and methionine were identified as helix formers, while proline and glycine, due to the unique conformational properties of their peptide bonds, commonly end a helix. The original Chou-Fasman parameters were derived from a very small and non-representative sample of protein structures due to the small number of such structures that were known at the time of their original work. These original parameters have since been shown to be unreliable and have been updated from a current dataset, along with modifications to the initial algorithm.

The Chou-Fasman method takes into account only the probability that each individual amino acid will appear in a helix, strand, or turn. Unlike the more complex GOR method, it does not reflect the conditional probabilities of an amino acid to form a particular secondary structure given that its neighbours already possess that structure. This lack of cooperativity increases its computational efficiency but decreases its accuracy, since the propensities of individual amino acids are often not strong enough to render a definitive prediction.

Algorithm

The Chou-Fasman method predicts helices and strands in a similar fashion, first searching linearly through the sequence for a "nucleation" region of high helix or strand probability and then extending the region until a subsequent four-residue window carries a probability of less than 1. As originally described, four out of any six contiguous amino acids were sufficient to nucleate helix, and three out of any contiguous five were sufficient for a sheet. The probability thresholds for helix and strand nucleations are constant but not necessarily equal; originally 1.03 was set as the helix cutoff and 1.00 for the strand cutoff.

Turns are also evaluated in four-residue windows, but are calculated using a multi-step procedure because many turn regions contain amino acids that could also appear in helix or sheet regions. Four-residue turns also have their own characteristic amino acids; proline and glycine are both common in turns. A turn is predicted only if the turn probability is greater than the helix or sheet probabilities *and* a probability value based on the positions of particular amino acids in the turn exceeds a predetermined threshold. The turn probability p(t) is determined as:

$$p(t) = p_t(j) \times p_t(j + 1) \times p_t(j + 2) \times p_t(j + 3)$$

where *j* is the position of the amino acid in the four-residue window. If p(t) exceeds an arbitrary cutoff value (originally 7.5e-3), the mean of the p(j)'s exceeds 1, and p(t) exceeds the alpha helix and beta sheet probabilities for that window, then a turn is predicted. If the first two conditions are met but the probability of a beta sheet p(b) exceeds p(t), then a sheet is predicted instead.

GOR Method

The GOR method (Garnier-Osguthorpe-Robson) is an information theory-based method for the prediction of secondary structures in proteins. It was developed in the late 1970's shortly after the simpler Chou-Fasman method. Like Chou-Fasman, the GOR method is based on probability parameters derived from empirical studies of known protein tertiary structures solved by X-ray crystallography. However, unlike Chou-Fasman, the GOR method takes into account not only the propensities of individual amino acids to form particular secondary structures, but also the conditional probability of the amino acid to form a secondary structure given that its immediate neighbours have already formed that structure. The method is therefore essentially Bayesian in its analysis.

The GOR method analyses sequences to predict alpha helix, beta sheet, turn, or random coil secondary structure at each position based on 17-amino acid sequence windows. The original description of the method included four scoring matrices of size 17x20, where the columns correspond to the log-odds score, which reflects the probability of finding a given amino acid at each position in the 17-residue sequence. The four matrices reflect the probabilities of the central, ninth amino acid being in a helical, sheet, turn, or coil conformation. In subsequent revisions to the method, the turn matrix was eliminated due to the high variability of sequences in turn regions (particularly over such a large window). The method was considered as best requiring at least four contiguous residues to score as alpha helices to classify the region as helical, and at least two contiguous residues for a beta sheet.

The mathematics and algorithm of the GOR method were based on a earlier series of studies by Robson and colleagues reported mainly in the Journal of Molecular Biology (e.g.) and The Biochemical Journal (e.g.). The latter describes the information theoretic expansions in terms of conditional information measures. The use of the word "simple" in the title of the GOR paper reflected the fact that the above earlier methods provided proofs and techniques somewhat daunting by being rather unfamilar in protein science in the early 1970s; even Bayes methods were then unfamilar and controversial.

An important feature of these early studies, which survived in the GOR method, was the treatment of the sparse protein sequence data of the early 1970s by expected information measures. That is, expectations on a Bayesian basis considering the distribution of plausible information measure values given the actual frequencies (numbers of observations). The expectation measures resulting from integration over this and similar distributions may now be seen as composed of "incomplete" or extended zeta functions, e.g. $z(s,\text{observed frequency}) - z(s,\text{expected frequency})$ with incomplete zeta function $z(s, n) = 1 + (1/2)^6 + (1/3)^s + (1/4)^s + \ldots + (1/n)^s$. The GOR method used $s=1$.

Also, in the GOR method and the earlier methods, the measure for the contrary state to e.g. helix H, i.e. ~H, was subtracted from that for H, and similarly for beta sheet, turns, and coil or loop. Thus the method can be seen as employing a zeta function estimate of log predictive odds. An adjustable decision constant could also be applied, which thus also implies a decision theory approach; the GOR method allowed the option to use decision constants to optimize predictions for different classes of protein. The expected information measure used as a basis for the information expansion was less important by the time of publication of the GOR method because protein sequence data became more plentiful, at least for the terms considered at that time. Then, for s=1, the expression z(s,observed frequency) – z(s,expected frequency) approaches the natural logarithm of (observed frequency/expected frequency) as frequencies increase. However, this measure (including use of other values of s) remains important in later more general applications with high dimensional data, where data for more complex terms in the information expansion are inevitably sparse (e.g.).

Machine Learning

Neural network methods use training sets of solved structures to identify common sequence motifs associated with particular arrangements of secondary structures. These methods are over 70% accurate in their predictions, although beta strands are still often underpredicted due to the lack of three-dimensional structural information that would allow assessment of hydrogen bonding patterns that can promote formation of the extended conformation required for the presence of a complete beta sheet.

Support vector machines have proven particularly useful for predicting the locations of turns, which are difficult to identify with statistical methods. The requirement of relatively small training sets has also been cited as an advantage to avoid overfitting to existing structural data.

Extensions of machine learning techniques attempt to predict more fine-grained local properties of proteins, such as backbone dihedral angles in unassigned regions. Both SVMs and neural networks have been applied to this problem.

Other Improvements

It is reported that in addition to the protein sequence, secondary structure formation depends on other factors. For example, it is reported that secondary structure tendencies depend also on local environment, solvent accessibility of residues, protein structural class, and even the organism from which the proteins are obtained. Based on such observations, some studies have shown that secondary structure prediction can be improved by addition of information about protein structural class, solvent accessibility and also contact number of residues. Sequence covariation methods rely on the existence of a data set composed of multiple homologous RNA sequences with related but dissimilar sequences. These methods analyse the covariation of individual base sites in evolution; maintenance at two widely separated sites of a pair of base-pairing nucleotides indicates the presence of a structurally required hydrogen bond between those positions. The general problem of pseudoknot prediction has been shown to be NP-complete.

Tertiary Structure

The practical role of protein structure prediction is now more important than ever. Massive amounts of protein sequence data are produced by modern large-scale DNA sequencing efforts such as the Human Genome Project. Despite community-wide efforts in structural genomics, the output of experimentally determined protein structures— typically by time-consuming and relatively expensive X-ray crystallography or NMR spectroscopy—is lagging far behind the output of protein sequences.

The protein structure prediction remains an extremely difficult and unresolved undertaking. The two main problems are calculation of protein free energy and finding the global minimum of this energy. A protein structure prediction method must explore the space of possible protein structures which is astronomically large.

These problems can be partially bypassed in "comparative" or homology modelling and fold recognition methods, when the search space is pruned by the assumption that the protein in question adopts a structure that is close to the experimentally determined structure of another homologous protein. On the other hand, the *de novo* or ab initio protein structure prediction methods must explicitly resolve these problems.

De Novo Protein Structure Prediction

In computational biology, *de novo* protein structure prediction is the task of estimating a protein's tertiary structure from its sequence alone. The problem is very difficult and has occupied leading scientists for decades. Research has focused in three areas: alternate lower-resolution representations of proteins, accurate energy functions, and efficient sampling methods. At present, the most successful methods have a reasonable probability of predicting the fold of a small protein domain within 5 angstroms.

De novo protein structure prediction methods attempt to predict tertiary structures from sequences based on general principles that govern protein folding energetics and/or statistical tendencies of conformational features that native structures acquire, without the use of explicit templates. A general paradigm for *de novo* prediction involves sampling conformation space, guided by scoring functions and other sequence-dependent biases such that a large set of candidate ("decoy") structures are generated. Native-like conformations are then selected from these decoys using scoring functions as well as conformer clustering. High-resolution refinement is sometimes used as a final step to fine-tune native-like structures. There are two major classes of scoring functions. Physics-based functions are based on mathematical models describing aspects of the known physics of molecular interaction. Knowledge-based functions are formed with statistical models capturing aspects of the properties of native protein conformations.

De novo methods tend to require vast computational resources, and have thus only been carried out for relatively small proteins. To predict protein structure *de novo* for larger proteins will require better algorithms and larger computational resources like those afforded by either powerful supercomputers (such as Blue Gene or MDGRAPE-3) or distributed computing projects (such as Rosetta@home, the Human Proteome Folding Project, or Nutritious Rice for the World). Although computational barriers are vast, the potential benefits of structural genomics (by predicted or experimental methods) make *de novo* structure prediction an active research field.

Comparative Protein Modelling

Comparative protein modelling uses previously solved structures as starting points, or templates. This is effective because it appears that although the number of actual proteins is vast, there is a limited set of tertiary structural motifs to which most proteins belong. It has been suggested that there are only around 2,000 distinct protein folds in nature, though there are many millions of different proteins.

Homology Modelling

Homology modelling, also known as comparative modelling of protein refers to constructing an atomic-resolution model of the *"target"* protein from its amino acid sequence and an experimental three-dimensional structure of a related homologous protein (the *"template"*). Homology modelling relies on the identification of one or more known protein structures likely to resemble the structure of the query sequence, and on the production of an alignment that maps residues in the query sequence to residues in the template sequence. It has been shown that protein structures are more conserved than protein sequences amongst homologues, but sequences falling below a 20% sequence identity can have very different structure.

Evolutionarily related proteins have similar sequences and naturally occurring homologous proteins have similar protein structure. It has been shown that three-dimensional protein structure is evolutionarily more conserved than expected due to sequence conservation.

The sequence alignment and template structure are then used to produce a structural model of the target. Because protein structures are more conserved than DNA sequences, detectable levels of sequence similarity usually imply significant structural similarity.

The quality of the homology model is dependent on the quality of the sequence alignment and template structure. The approach can be complicated by the presence of alignment gaps (commonly called indels) that indicate a structural region present in the target but not in the template, and by structure gaps in the template that arise from poor resolution in the experimental procedure (usually X-ray crystallography) used to solve the structure. Model quality declines with decreasing sequence identity; a typical model has ~1-2 Å root mean square deviation between the matched C atoms at 70% sequence identity but only 2-4 Å agreement at 25% sequence identity. However, the errors are significantly higher in the loop regions, where the amino acid sequences of the target and template proteins may be completely different.

Regions of the model that were constructed without a template, usually by loop modelling, are generally much less accurate than the rest of the model. Errors in side chain packing and position also increase with decreasing identity, and variations in these packing configurations have been suggested as a major reason for poor model quality at low identity. Taken together, these various atomic-position errors are significant and impede the use of homology models for purposes that require atomic-resolution data, such as drug design and protein-protein interaction predictions; even the quaternary structure of a protein may be difficult to predict from homology models of its subunit(s). Nevertheless, homology models can be useful in reaching *qualitative* conclusions about the biochemistry of the query sequence, especially in formulating hypotheses about why certain residues are conserved, which may in turn lead to experiments to test those hypotheses. For example, the spatial arrangement of conserved residues may suggest whether a particular residue is conserved to stabilize the folding, to participate in binding some small molecule, or to foster association with another protein or nucleic acid.

Homology modelling can produce high-quality structural models when the target and template are closely related, which has inspired the formation of a structural genomics consortium dedicated to the production of representative experimental structures for all classes of protein folds. The chief inaccuracies in homology modelling, which worsen with lower sequence identity, derive from errors in the initial sequence alignment and from improper template selection. Like other methods of structure prediction, current practice in homology modelling is assessed in a biannual large-scale experiment known as the Critical Assessment of Techniques for Protein Structure Prediction, or CASP.

Motive

The method of homology modelling is based on the observation that protein tertiary structure is better conserved than amino acid sequence. Thus, even proteins that have diverged appreciably in sequence but still share detectable similarity will also share common structural properties, particularly the overall fold. Because it is difficult and time-consuming to obtain experimental structures from methods such as X-ray crystallography and protein NMR for every protein of interest, homology modelling can provide useful structural models for generating hypotheses about a protein's function and directing further experimental work.

There are exceptions to the general rule that proteins sharing significant sequence identity will share a fold. For example, a judiciously chosen set of mutations of less than 50% of a protein can cause the protein to adopt a completely different fold. However, such a massive structural rearrangement is unlikely to occur in evolution, especially since the protein is usually under the constraint that it must fold properly and carry out its function in the cell. Consequently, the roughly folded structure of a protein (its "topology") is conserved longer than its amino-acid sequence and much longer than the corresponding DNA sequence; in other words, two proteins may share a similar fold even if their evolutionary relationship is so distant that it cannot be discerned reliably. For comparison, the function of a protein is conserved much *less* than the protein sequence, since relatively few changes in amino-acid sequence are required to take on a related function.

Steps in Model Production

The homology modelling procedure can be broken down into four sequential steps: template selection, target-template alignment, model construction, and model assessment. The first two steps are often essentially performed together, as the most common methods of identifying templates rely on the production of sequence alignments; however, these alignments may not be of sufficient quality because database search techniques prioritize speed over alignment quality. These processes can be performed iteratively to improve the quality of the final model, although quality assessments that are not dependent on the true target structure are still under development.

Optimizing the speed and accuracy of these steps for use in large-scale automated structure prediction is a key component of structural genomics initiatives, partly because the resulting volume of data will be too large to process manually and partly because the goal of structural genomics requires providing models of reasonable quality to researchers who are not themselves structure prediction experts.

Template Selection and Sequence Alignment

The critical first step in homology modelling is the identification of the best template structure, if indeed any are available. The simplest method of template identification relies on serial pairwise sequence alignments aided by database search techniques such as FASTA and BLAST. More sensitive methods based on multiple sequence alignment-of which PSI-BLAST is the most common example-iteratively update their position-specific scoring matrix to successively identify more distantly related homologs. This family of methods has been shown to produce a larger number of potential templates and to identify better templates for sequences that have only distant relationships to any solved structure. Protein threading, also known as fold recognition or 3D-1D alignment, can also be used as a search technique for identifying templates to be used in traditional homology modelling methods. When performing a BLAST search, a reliable first approach is to identify hits with a sufficiently low *E*-value, which are considered sufficiently close in evolution to make a reliable homology model. Other factors may tip the balance in marginal cases; for example, the template may have a function similar to that of the query sequence, or it may belong to a homologous operon. However, a template with a poor *E*-value should generally not be chosen, even if it is the only one available, since it may well have a wrong structure, leading to the production of a misguided model. A better approach is to submit the primary sequence to fold-recognition servers or, better still, consensus meta-servers which improve upon individual fold-recognition servers by identifying similarities (consensus) among independent predictions.

Often several candidate template structures are identified by these approaches. Although some methods can generate hybrid models from multiple templates, most methods rely on a single template. Therefore, choosing the best template from among the candidates is a key step, and can affect the final accuracy of the structure significantly. This choice is guided by several factors, such as the similarity of the query and template sequences, of their functions, and of the predicted query and observed template secondary structures. Perhaps most importantly, the *coverage* of the aligned regions: the fraction of the query sequence structure that can be predicted from the template, and the plausibility of the resulting model. Thus, sometimes several homology models are produced for a single query sequence, with the most likely candidate chosen only in the final step.

It is possible to use the sequence alignment generated by the database search technique as the basis for the subsequent model production; however, more sophisticated approaches have also been explored. One proposal generates an ensemble of stochastically defined pairwise alignments between the target sequence and a single identified template as a means of exploring "alignment space" in regions of sequence with low local similarity. "Profile-profile" alignments that first generate a sequence profile of the target and systematically compare it to the sequence profiles of solved structures; the coarse-graining inherent in the profile construction is thought to reduce noise introduced by sequence drift in nonessential regions of the sequence.

Model Generation

Given a template and an alignment, the information contained therein must be used to generate a three-dimensional structural model of the target, represented as a set of Cartesian coordinates for each atom in the protein. Three major classes of model generation methods have been proposed.

Fragment Assembly

The original method of homology modelling relied on the assembly of a complete model from conserved structural fragments identified in closely related solved structures. For example, a modelling study of serine proteases in mammals identified a sharp distinction between "core" structural regions conserved in all experimental structures in the class, and variable regions typically located in the loops where the majority of the sequence differences were localized. Thus unsolved proteins could be modelled by first constructing the conserved core and then substituting variable regions from other proteins in the set of solved structures. Current implementations of this method differ mainly in the way they deal with regions that are not conserved or that lack a template. The variable regions are often constructed with the help of fragment libraries.

Segment Matching

The segment-matching method divides the target into a series of short segments, each of which is matched to its own template fitted from the Protein Data Bank. Thus, sequence alignment is done over segments rather than over the entire protein. Selection of the template for each segment is based on sequence similarity, comparisons of alpha carbon coordinates, and predicted steric conflicts arising from the van der Waals radii of the divergent atoms between target and template.

Satisfaction of Spatial Restraints

The most common current homology modelling method takes its inspiration from calculations required to construct a three-dimensional structure from data generated by NMR spectroscopy. One or more target-template alignments are used to construct a set of geometrical criteria that are then converted to probability density functions for each restraint. Restraints applied to the main protein internal coordinates-protein backbone distances and dihedral angles-serve as the basis for a global optimization procedure that originally used conjugate gradient energy minimization to iteratively refine the positions of all heavy atoms in the protein.

This method had been dramatically expanded to apply specifically to loop modelling, which can be extremely difficult due to the high flexibility of loops in proteins in aqueous solution. A more recent expansion applies the spatial-restraint model to electron density maps derived from cryoelectron microscopy studies, which provide low-resolution information that is not usually itself sufficient to generate atomic-resolution structural models.

To address the problem of inaccuracies in initial target-template sequence alignment, an iterative procedure has also been introduced to refine the alignment on the basis of the initial structural fit. The most commonly used software in spatial restraint-based modelling is MODELLER and a database called ModBase has been established for reliable models generated with it.

Loop Modelling

Regions of the target sequence that are not aligned to a template are modelled by loop modelling; they are the most susceptible to major modelling errors and occur with higher frequency when the target and template have low sequence identity. The coordinates of unmatched sections determined by loop modelling programs are generally much less accurate than those obtained from simply copying the coordinates of a known structure, particularly if the loop is longer than 10 residues. The first two sidechain dihedral angles ($\div_1$ and $\div_2$) can usually be estimated within 30° for an accurate backbone structure; however, the later dihedral angles found in longer side chains such as lysine and arginine are notoriously difficult to predict. Moreover, small errors in $\div_1$ (and, to a lesser extent, in $\div_2$) can cause relatively large errors in the positions of the atoms at the terminus of side chain; such atoms often have a functional importance, particularly when located near the active site.

Model Assessment

Assessment of homology models without reference to the true target structure is usually performed with two methods: statistical potentials or physics-based energy calculations. Both methods produce an estimate of the energy (or an energy-like analog) for the model or models being assessed; independent criteria are needed to determine acceptable cutoffs. Neither of the two methods correlates exceptionally well with true structural accuracy, especially on protein types underrepresented in the PDB, such as membrane proteins.

Statistical potentials are empirical methods based on observed residue-residue contact frequencies among proteins of known structure in the PDB. They assign a probability or energy score to each possible pairwise interaction between amino acids and combine these pairwise interaction scores into a single score for the entire model. Some such methods can also produce a residue-by-residue assessment that identifies poorly scoring regions within the model, though the model may have a reasonable score overall. These methods emphasize the hydrophobic core and solvent-exposed polar amino acids often present in globular proteins. Examples of popular statistical potentials include Prosa and DOPE. Statistical potentials are more computationally efficient than energy calculations.

Physics-based energy calculations aim to capture the interatomic interactions that are physically responsible for protein stability in solution, especially van der Waals and electrostatic interactions. These calculations are performed using a molecular mechanics force field; proteins are normally too large even for semi-empirical quantum mechanics-based calculations. The use of these methods is based on the energy landscape hypothesis of protein folding, which predicts that a protein's native state is also its energy minimum. Such methods usually employ implicit solvation, which provides a continuous approximation of a solvent bath for a single protein molecule without necessitating the explicit representation of individual solvent molecules. A force field specifically constructed for model assessment is known as the Effective Force Field (EFF) and is based on atomic parameters from CHARMM.

A very extensive model validation report can be obtained using the Radboud Universiteit Nijmegen *What Checkf* software which is one option of the Radboud Universiteit Nijmegen *What Iff* software package; it produces a many page document with extensive analyses of nearly 200 scientific and administrative aspects of the model. *What Checkf* is available as a free server; it can also be used to validate experimentally determined structures of macromolecules.

One newer method for model assessment relies on machine learning techniques such as neural nets, which may be trained to assess the structure directly or to form a consensus among multiple statistical and energy-based methods. Very recent results using support vector machine regression on a jury of more traditional assessment methods outperformed common statistical, energy-based, and machine learning methods.

Structural Comparison Methods

The assessment of homology models' accuracy is straightforward when the experimental structure is known. The most common method of comparing two protein structures uses the root-mean-square deviation (RMSD) metric to measure the mean distance between the corresponding atoms in the two structures after they have been superimposed. However, RMSD does underestimate the accuracy of models in which the core is essentially correctly modelled, but some flexible loop regions are inaccurate. A method introduced for the modelling assessment experiment CASP is known as the global distance test (GDT) and measures the total number of atoms whose distance from the model to the experimental structure lies under a certain distance cutoff. Both methods can be used for any subset of atoms in the structure, but are often applied to only the alpha carbon or protein backbone atoms to minimize the noise created by poorly modelled side chain rotameric states, which most modelling methods are not optimized to predict.

Benchmarking

Several large-scale benchmarking efforts have been made to assess the relative quality of various current homology modelling methods. CASP is a community-wide prediction experiment that runs every two years during the summer months and challenges prediction teams to submit structural models for a number of sequences whose structures have recently been solved experimentally but have not yet been published. Its partner CAFASP has run in parallel with CASP but evaluates only models produced via fully automated servers. Continuously running experiments that do not have prediction 'seasons' focus mainly on benchmarking publicly available webservers. LiveBench and EVA run continuously to assess participating servers' performance in prediction of imminently released structures from the PDB. CASP and CAFASP serve mainly as evaluations of the state of the art in modelling, while the continuous assessments seek to evaluate the model quality that would be obtained by a non-expert user employing publicly available tools.

Accuracy

The accuracy of the structures generated by homology modelling is highly dependent on the sequence identity between target and template. Above 50% sequence identity, models tend to be reliable, with only minor errors in side chain packing and rotameric state, and an overall RMSD between the modelled and the experimental structure falling around 1 Å. This error is comparable to the typical resolution of a structure solved by NMR. In the 30-50% identity range, errors can be more severe and are often located in loops. Below 30% identity, serious errors occur, sometimes resulting in the basic fold being mispredicted. This low-identity region is often referred to as the "twilight zone" within which homology modelling is extremely difficult, and to which it is possibly less suited than fold recognition methods. At high sequence identities, the primary source of error in homology modelling derives from the choice of the template or templates on which the model is based, while lower identities exhibit serious errors in sequence alignment that inhibit the production of high-quality models. It has been suggested that the major impediment to quality model production is inadequacies in sequence alignment, since "optimal" structural alignments between two proteins of known structure can be used as input to current modelling methods to produce quite accurate reproductions of the original experimental structure.

Attempts have been made to improve the accuracy of homology models built with existing methods by subjecting them to molecular dynamics simulation in an effort to improve their RMSD to the experimental structure. However, current force field parameterizations may not be sufficiently accurate for this task, since homology models used as starting structures for molecular dynamics tend to produce slightly worse structures. Slight improvements have been observed in cases where significant restraints were used during the simulation.

Sources of Error

The two most common and large-scale sources of error in homology modelling are poor template selection and inaccuracies in target-template sequence alignment. Controlling for these two factors by using a structural alignment, or a sequence alignment produced on the basis of comparing two solved structures, dramatically reduces the errors in final models; these "gold standard" alignments can be used as input to current modelling methods to produce quite accurate reproductions of the original experimental structure. Results from the most recent CASP experiment suggest that "consensus" methods collecting the results of multiple fold recognition and multiple alignment searches increase the likelihood of identifying the correct template; similarly, the use of multiple templates in the model-building step may be less optimal than the use of the single correct template but more optimal than the use of a single suboptimal one. Alignment errors may be minimized by the use of a multiple alignment even if only one template is used, and by the iterative refinement of local regions of low similarity. A lesser source of model errors are errors in the template structure. The PDBREPORT database lists several million, mostly very small but occasionally dramatic, errors in experimental (template) structures that have been deposited in the PDB. Serious local errors can arise in homology models where an insertion or deletion mutation or a gap in a solved structure result in a region of target sequence for which there is no corresponding template. This problem can be minimized by the use of multiple templates, but the method is complicated by the templates' differing local structures around the gap and by the likelihood that a missing region in one experimental structure is also missing in other structures of the same protein family.

Missing regions are most common in loops where high local flexibility increases the difficulty of resolving the region by structure-determination methods. Although some guidance is provided even with a single template by the positioning of the ends of the missing region, the longer the gap, the more difficult it is to model. Loops of up to about 9 residues can be modelled with moderate accuracy in some cases if the local alignment is correct. Larger regions are often modelled individually using ab initio structure prediction techniques, although this approach has met with only isolated success.

The rotameric states of side chains and their internal packing arrangement also present difficulties in homology modelling, even in targets for which the backbone structure is relatively easy to predict. This is partly due to the fact that many side chains in crystal structures are not in their "optimal" rotameric state as a result of energetic factors in the hydrophobic core and in the packing of the individual molecules in a protein crystal. One method of addressing this problem requires searching a rotameric library to identify locally low-energy combinations of packing states. It has been suggested that a major reason that homology modelling so difficult when target-template sequence identity lies below 30% is that such proteins have broadly similar folds but widely divergent side chain packing arrangements.

Utility

Uses of the structural models include protein-protein interaction prediction, protein-protein docking, molecular docking, and functional annotation of genes identified in an organism's genome. Even low-accuracy homology models can be useful for these purposes, because their inaccuracies tend to be located in the loops on the protein surface, which are normally more variable even between closely related proteins. The functional regions of the protein, especially its active site, tend to be more highly conserved and thus more accurately modelled.

Homology models can also be used to identify subtle differences between related proteins that have not all been solved structurally. For example, the method was used to identify cation binding sites on the Na⁺/K⁺ ATPase and to propose hypotheses about different ATPases' binding affinity. Used in conjunction with molecular dynamics simulations, homology models can also generate hypotheses about the kinetics and dynamics of a protein, as in studies of the ion selectivity of a potassium channel. Large-scale automated modelling of all identified protein-coding regions in a genome has been attempted for the yeast *Saccharomyces cerevisiae*, resulting in nearly 1000 quality models for proteins whose structures had not yet been determined at the time of the study, and identifying novel relationships between 236 yeast proteins and other previously solved structures.

Threading (Protein Sequence)

Protein threading, also known as fold recognition, is a method of protein modelling (i.e. computational protein structure prediction) which is used to model those proteins which have the same fold as proteins of known structures, but do not have homologous proteins with known structure. It differs from the homology modelling method of structure prediction in that it is used for proteins which do not have their homologous protein structures deposited in the Protein Data Bank (PDB), whereas homology modelling is used for those proteins which do. Threading works by using statistical knowledge of the relationship between the structures deposited in the PDB and the sequence of the protein which one wishes to model.

The prediction is made by "threading" (i.e. placing, aligning) each amino acid in the target sequence to a position in the template structure, and evaluating how well the target fits the template. After the best-fit template is selected, the structural model of the sequence is built based on the alignment with the chosen template. Protein threading is based on two basic observations: that the number of different folds in nature is fairly small (approximately 1000); and that 90% of the new structures submitted to the PDB in the past three years have similar structural folds to ones already in the PDB.

Classification of Protein Structure

The Structural Classification of Proteins (SCOP) database provides a detailed and comprehensive description of the structural and evolutionary relationships of known structure. Proteins are classified to reflect both structural and evolutionary relatedness. Many levels exist in the hierarchy, but the principal levels are family, superfamily and fold, as described below.

Family (clear evolutionary relationship) : Proteins clustered together into families are clearly evolutionarily related. Generally, this means that pairwise residue identities between the proteins are 30% and greater. However, in some cases similar functions and structures provide definitive evidence of common descent in the absence of high sequence identity; for example, many globins form a family though some members have sequence identities of only 15%.

Superfamily (probable common evolutionary origin : Proteins that have low sequence identities, but whose structural and functional features suggest that a common evolutionary origin is probable, are placed together in superfamilies. For example, actin, the ATPase domain of the heat shock protein, and hexakinase together form a superfamily.

Fold (major structural similarity): Proteins are defined as having a common fold if they have the same major secondary structures in the same arrangement and with the same topological connections. Different proteins with the same fold often have peripheral elements of secondary structure and turn regions that differ in size and conformation. In some cases, these differing peripheral regions may comprise half the structure. Proteins placed together in the same fold category may not have a common evolutionary origin: the structural similarities could arise just from the physics and chemistry of proteins favouring certain packing arrangements and chain topologies.

Method

A general paradigm of protein threading consists of the following four steps:

The construction of a structure template database : Select protein structures from the protein structure databases as structural templates. This generally involves selecting protein structures from databases such as PDB, FSSP, SCOP, or CATH, after removing protein structures with high sequence similarities.

The design of the scoring function : Design a good scoring function to measure the fitness between target sequences and templates based on the knowledge of the known relationships between the structures and the sequences. A good scoring function should contain mutation potential, environment fitness potential, pairwise potential, secondary structure compatibilities, and gap penalties. The quality of the energy function is closely related to the prediction accuracy, especially the alignment accuracy.

Threading alignmen : Align the target sequence with each of the structure templates by optimizing the designed scoring function. This step is one of the major tasks of all threading-based structure prediction programs that take into account the pairwise contact potential; otherwise, a dynamic programming algorithm can fulfil it. This thesis is mainly dedicated to solving the optimal alignment problem derived from a scoring function considering pairwise contacts.

Threading prediction: Select the threading alignment that is statistically most probable as the threading prediction. Then construct a structure model for the target by placing the backbone atoms of the target sequence at their aligned backbone positions of the selected structural template.

Comparison with Homology Modelling

Homology modelling and protein threading are both template-based methods and there is no rigorous boundary between them in terms of prediction techniques. But the protein structures they target are different. Homology modelling is for those targets that have homologous proteins with known structure, while protein threading is for those targets with only fold-level homology found. In other words, homology modelling is for "easier" targets and protein threading is for "harder" targets.

Homology modelling treats the template in an alignment as a sequence, and only sequence homology is used for prediction. Protein threading treats the template in an alignment as a structure, and both sequence and structure information extracted from the alignment are used for prediction. When there is no significant homology found, protein threading can make a prediction based on the structure information. That also explains why protein threading may be more effective than homology modelling in many cases.

In practice, when the sequence identity in a sequence sequence alignment is low (i.e. <25%), homology modelling may not produce a significant prediction. In this case, if there is distant homology found for the target, protein threading can generate a good prediction.

More about Threading

Fold recognition methods can be broadly divided into two types: 1, those that derive a 1-D profile for each structure in the fold library and align the target sequence to these profiles; and 2, those that consider the full 3-D structure of the protein template. A simple example of a profile representation would be to take each amino acid in the structure and simply label it according to whether it is buried in the core of the protein or exposed on the surface.

More elaborate profiles might take into account the local secondary structure (e.g. whether the amino acid is part of an alpha helix) or even evolutionary information (how conserved the amino acid is). In the 3-D representation, the structure is modelled as a set of inter-atomic distances, i.e. the distances are calculated between some or all of the atom pairs in the structure. This is a much richer and far more flexible description of the structure, but is much harder to use in calculating an alignment. The profile-based fold recognition approach was first described by Bowie, Luthy and Eisenberg in 1991. The term *threading* was first coined by Jones, Taylor and Thornton in 1992, and originally referred specifically to the use of a full 3-D structure atomic representation of the protein template in fold recognition. Today, the terms threading and fold recognition are frequently (though somewhat incorrectly) used interchangeably.

Fold recognition methods are widely used and effective because it is believed that there are a strictly limited number of different protein folds in nature, mostly as a result of evolution but also due to constraints imposed by the basic physics and chemistry of polypeptide chains. There is, therefore, a good chance (currently 70-80%) that a protein which has a similar fold to the target protein has already been studied by X-ray crystallography or nuclear magnetic resonance (NMR) spectroscopy and can be found in the PDB. Currently there are just over 1100 different protein folds known, but new folds are still being discovered every year due in significant part to the ongoing structural genomics projects.

Many different algorithms have been proposed for finding the correct threading of a sequence onto a structure, though many make use of dynamic programming in some form. For full 3-D threading, the problem of identifying the best alignment is very difficult (it is an NP-hard problem for some models of threading). Researchers have made use of many combinatorial optimization methods such as simulated annealing or branch and bound searching to arrive at heuristic solutions.

It is interesting to compare threading methods to methods which attempt to align two protein structures (protein structural alignment), and indeed many of the same algorithms have been applied to both problems.

Protein Threading Software

- HHpred is a popular threading server which runs HHsearch, a widely used software for remote homology detection based on pairwise comparison of hidden Markov models.

- RAPTOR (software) is an integer programming based protein threading software.

Side Chain Geometry Prediction

Accurate packing of the amino acid side chains represents a separate problem. Methods that specifically address the problem of predicting side chain geometry include dead-end elimination and the self-consistent mean field methods. The side chain conformations with low energy are usually determined on the rigid polypeptide backbone and using a set of discrete side chain conformations known as "rotamers" or a "conformational isomerism". The methods attempt to identify the set of rotamers that minimize the model's overall energy.

These methods use rotamer libraries, the collections of rotamers (favourable multi-angle conformations) for each residue type in proteins. Rotamer libraries may contain information about the conformation, its frequency, and the variance about mean dihedral angles, which can be used in sampling. Rotamer libraries are derived from structural bioinformatics or other statistical analysis of side-chain conformations in known experimental structures of proteins, such as by clustering the observed conformations for tetrahedral carbons near the staggered (60°, 180°, -60°) values.

Rotamer libraries can be backbone-independent, secondary-structure-dependent, or backbone-dependent. Backbone-independent rotamer libraries make no reference to backbone conformation, and are calculated from all available side chains of a certain type (for instance, the first example of a rotamer library, done by Ponder and Richards at Yale in 1987). Secondary-structure-dependent libraries present different dihedral angles and/or rotamer frequencies for α-helix, β-sheet, or coil secondary structures. Backbone-dependent rotamer libraries present conformations and/or frequencies dependent on the local backbone conformation as defined by the backbone dihedral angles ϕ and ψ, regardless of secondary structure. The modern versions of these "libraries" as used in most software are presented as multidimensional distributions of probability or frequency, where the peaks correspond to the dihedral-angle conformations considered as individual rotamers in the lists. Some versions are especially sensitive to the prohibited regions in that conformational space and are used primarily for structure validation, while others emphasize relative frequencies in the favourable regions and are the form used primarily for structure prediction, such as the Dunbrack rotamer "libraries".

The side chain packing methods are most useful for analysing the protein's hydrophobic core, where side chains are more closely packed; they have more difficulty addressing the looser constraints and higher flexibility of surface residues, which often occupy multiple rotamer conformations rather than just one.

Prediction of Structural Classes

Statistical methods have been developed for predicting structural classes of proteins based on their amino acid composition, pseudo amino acid composition and functional domain composition.

Protein˜protein Interaction Prediction

Protein–protein interaction prediction is a field combining bioinformatics and structural biology in an attempt to identify and catalogue physical interactions between pairs or groups of proteins. Understanding protein–protein interactions is important for the investigation of intracellular signalling pathways, modelling of protein complex structures and for gaining insights into various biochemical processes. Experimentally, physical interactions between pairs of proteins can be inferred from a variety of experimental techniques, including yeast two-hybrid systems, protein-fragment complementation assays (PCA), affinity purification/mass spectrometry, protein microarrays and fluorescence resonance energy transfer (FRET). Efforts to experimentally determine the interactome of numerous species are ongoing, and a number of computational methods for interaction prediction have been developed in recent years.

Methods

Proteins that interact are more likely to co-evolve, therefore it is possible to make inferences about interactions between pairs of proteins based on their phylogenetic distances. It has also been observed in some cases that pairs of interacting proteins have fused orthologues in other organisms. In addition, a number of bound protein complexes have been structurally solved and can be used to identify the residues that mediate the interaction so that similar motifs can be located in other organisms.

Phylogenetic Profiling

Phylogenetic profiling finds pairs of protein families with similar patterns of presence or absence across large numbers of species. This method identifies pairs likely to act in the same biological process, but does not necessarily imply physical interaction.

Prediction of Co-evolved Protein Pairs based on Similar Phylogenetic Trees

This method involves using a sequence search tool such as BLAST for finding homologues of a pair of proteins, then building multiple sequence alignments with alignment tools such as Clustal. From these multiple sequence alignments, phylogenetic distance matrices are calculated for each protein in the hypothesized interacting pair. If the matrices are sufficiently similar (as measured by their Pearson correlation coefficient) they are deemed likely to interact.

Identification of Homologous Interacting Pairs

This method consists of searching whether the two sequences have homologues which form a complex in a database of known structures of complexes. The identification of the domains is done by sequence searches against domain databases such as Pfam using BLAST. If more than one complex of Pfam domains is identified, then the query sequences are aligned using a hidden Markov tool called HMMER to the closest identified homologues, whose structures are known. Then the alignments are analysed to check whether the contact residues of the known complex are conserved in the alignment.

Identification of Structural Patterns

This method builds a library of known protein–protein interfaces from the PDB, where the interfaces are defined as pairs of polypeptide fragments that are below a threshold slightly larger than the Van der Waals radius of the atoms involved. The sequences in the library are then clustered based on structural alignment and redundant sequences are eliminated. The residues that have a high (generally >50%) level of frequency for a given position are considered hotspots. This library is then used to identify potential interactions between pairs of targets, providing that they have a known structure (i.e. present in the PDB).

Bayesian Network Modelling

Bayesian methods integrate data from a wide variety of sources, including both experimental results and prior computational predictions, and use these features to assess the likelihood that a particular potential protein interaction is a true positive result. These methods are useful because experimental procedures, particularly the yeast two-hybrid experiments, are extremely noisy and produce many false positives, while the previously mentioned computational methods can only provide circumstantial evidence that a particular pair of proteins might interact.

3D Template-based Protein Complex Modelling

This method makes use of known protein complex structures to predict as well as structurally model interactions between query protein sequences. The prediction process generally starts by employing a sequence based method (e.g Interolog) to search for protein complex structures that are homologous to the query sequences.

These known complex structures are then used as templates to structurally model the interaction between query sequences. This method has the advantage of not only inferring protein interactions but also suggests models of how proteins interact structurally, which can provide some insights into the atomic level mechanism of that interaction. On the other hand, the ability for this method to makes a prediction is limited to a relatively small number of known protein complex structures.

Supervised Learning Problem

The problem of PPI prediction can be framed as a supervised learning problem. In this paradigm the known protein interactions supervise the estimation of a function that can predict whether an interaction exists or not between two proteins given data about the proteins (e.g., expression levels of each gene in different experimental conditions, location information, phylogenetic profile, etc.).

Relationship to Docking Methods

The field of protein–protein interaction prediction is closely related to the field of protein–protein docking, which attempts to use geometric and steric considerations to fit two proteins of known structure into a bound complex. This is a useful mode of inquiry in cases where both proteins in the pair have known structures and are known (or at least strongly suspected) to interact, but since so many proteins do not have experimentally determined structures, sequence-based interaction prediction methods are especially useful in conjunction with experimental studies of an organism's interactome.

Software

Modeller is a popular software tool for producing homology models using methodology derived from NMR spectroscopy data processing. SwissModel provides an automated web server for basic homology modelling.

I-tasser is the best server for protein structure prediction according to the 2006-2008 CASP experiments (CASP7 and CASP8).

HHpred, bioinfo.pl, Robetta, and Phyre are widely used servers for protein structure prediction. HHsearch is a free software package for protein threading and remote homology detection.

Raptor (software) is a protein threading software that is based on integer programming. The basic algorithm for threading is described in and is fairly straightforward to implement.

Abalone is a Molecular Dynamics program for folding simulations with explicit or implicit water models.

TIP is a knowledgebase of STRUCTFAST models and precomputed similarity relationships between sequences, structures, and binding sites. Several distributed computing projects concerning protein structure prediction have also been implemented, such as the Folding@home, Rosetta@home, Human Proteome Folding Project, Predictor@home, and TANPAKU.

The Foldit program seeks to investigate the pattern-recognition and puzzle-solving abilities inherent to the human mind in order to create more successful computer protein structure prediction software.

Computational approaches provide a fast alternative route to antibody structure prediction. Recently developed antibody F_V region high resolution structure prediction algorithms, like RosettaAntibody, have been shown to generate high resolution homology models which have been used for successful docking.

Reviews of software for structure prediction can be found at. The progress and challenges in protein structure prediction has been reviewed in Zhang 2008.

CASP

CASP, which stands for Critical Assessment of Techniques for Protein Structure Prediction, is a community-wide, worldwide experiment for protein structure prediction taking place every two years since 1994. CASP provides research groups with an opportunity to objectively test their structure prediction methods and delivers an independent assessment of the state of the art in protein structure modelling to the research community and software users. Even though the primary goal of CASP is to help advance the methods of identifying protein three-dimensional structure from its amino acid sequence, many view the experiment more as a "world championship" in this field of science. More than 100 research groups from all over the world participate in CASP on the regular basis and it is not uncommon for the entire groups to suspend their other research for months while they focus on getting their servers ready for the experiment and on performing the detailed predictions.

Selection of Target Proteins

In order to ensure that no predictor can have prior information about a protein's structure that would put him/her at an advantage, it is important that the experiment is conducted in a double-blind fashion: Neither predictors nor the organizers and assessors know the structures of the target proteins at the time when predictions are made. Targets for structure prediction are either structures soon-to-be solved by X-ray crystallography or NMR spectroscopy, or structures that have just been solved (mainly by one of the structural genomics centres) and are kept on hold by the Protein Data Bank. If the given sequence is found to be related by common descent to a protein sequence of known structure (called a template), comparative protein modelling may be used to predict the tertiary structure. Templates can be found using sequence alignment methods such as BLAST or FASTA or protein threading methods, which are better in finding distantly related templates. Otherwise, *de novo* protein structure prediction must be applied, which is much less reliable but can sometimes yield models with the correct fold. Truly new folds are becoming quite rare among the targets, making that category smaller than desirable.

Evaluation

The primary method of evaluation is a comparison of the predicted model α-carbon positions with those in the target structure. The comparison is shown visually by cumulative plots of distances between pairs of equivalents α-carbon in the alignment of the model and the structure, such as shown in the figure (a perfect model would stay at zero all the way across), and is assigned a numerical score GDT-TS (Global Distance Test-Total Score) describing percentage of well-modelled residues in the model with respect to the target. Free modelling (template-free, or *de novo*) is also evaluated visually by the assessors, since the numerical scores do not work as well for finding loose resemblances in the most difficult cases. High-accuracy template-based predictions were evaluated in CASP7 by whether they worked for molecular-replacement phasing of the target crystal structure with successes followed up later, and by full-model (not just α-carbon) model quality and full-model match to the target in CASP8.

Evaluation of the results is carried out in the following prediction categories:

- tertiary structure prediction (all CASPs)
- secondary structure prediction (dropped after CASP5)
- prediction of structure complexes (CASP2 only; a separate experiment-CAPRI-carries on this subject)
- residue-residue contact prediction (starting CASP4)
- disordered regions prediction (starting CASP5)
- domain boundary prediction (CASP6-CASP8)
- function prediction (starting CASP6)
- model quality assessment (starting CASP7)
- model refinement (starting CASP7)
- high-accuracy template-based prediction (starting CASP7).

Tertiary structure prediction category was further subdivided into;

- homology modelling
- fold recognition (also called protein threading; Note, this is incorrect as threading is a method)
- *de novo* structure prediction, now referred to as 'New Fold' as many methods apply evaluation, or scoring, functions that are biased by knowledge of native protein structures, such as an artificial neural network.

Starting with CASP7, categories have been redefined to reflect developments in methods. The 'Template based modeling' category includes all former comparative modelling, homologous fold based models and some analogous fold based models. The 'Template free modeling' category includes models of proteins with previously unseen folds and hard analogous fold based models.

The CASP results are published in special supplement issues of the scientific journal *Proteins*, all of which are accessible through the CASP website. A lead article in each of these supplements describes specifics of the experiment while a closing article evaluates progress in the field.

Result Ranking

Automated assessments for CASP9 (2010);

- Official ranking for servers only (147 targets)
- Official ranking for humans and servers (78 targets)
- Official ranking for free modelling targets (30 targets)
- Ranking by Grishin Lab (for server only)
- Ranking by Grishin Lab (for human and servers)
- Ranking by Zhang Lab
- Ranking by Cheng Lab.

Automated assessments for CASP8 (2008);

- Official ranking for servers only
- Official ranking for humans and servers
- Ranking by Zhang Lab
- Ranking by Grishin Lab
- Ranking McGuffin Lab
- Ranking by Cheng Lab.

Automated assessments for CASP7 (2006);

- Ranking by Livebench
- Ranking by Zhang Lab.

Human Biological Systems

Brain Model

The Blue Brain Project is an attempt to create a synthetic brain by reverse-engineering the mammalian brain down to the molecular level. The aim of the project, founded in May 2005 by the Brain and Mind Institute of the *École Polytechnique* in Lausanne, Switzerland, is to study the brain's architectural and functional principles. The project is headed by the Institute's director, Henry Markram. Using a Blue Gene supercomputer running Michael Hines's NEURON software, the simulation does not consist simply of an artificial neural network, but involves a partially biologically realistic model of neurons.

It is hoped by its proponents that it will eventually shed light on the nature of consciousness. There are a number of sub-projects, including the Cajal Blue Brain, coordinated by the Supercomputing and Visualization Centre of Madrid (CeSViMa), and others run by universities and independent laboratories in the UK, U.S., and Israel.

Model of the Immune System

The last decade has seen the emergence of a growing number of simulations of the immune system.

Simulated Growth of Plants

The simulated growth of plants is a significant task in of systems biology and mathematical biology, which seeks to reproduce plant morphology with computer software. Electronic trees (e-trees) usually use L-systems to simulate growth. L-systems are very important in the field of complexity science and A-life. A universally accepted system for describing changes in plant morphology at the cellular or modular level has yet to be devised. The most widely implemented tree-generating algorithms are described in the papers "Creation and Rendering of Realistic Trees", and Real-Time Tree Rendering

The realistic modelling of plant growth is of high value to biology, but also for computer games.

Theory + Algorithms

A biologist, Aristid Lindenmayer (1925–1989) worked with yeast and filamentous fungi and studied the growth patterns of various types of algae, such as the blue/green bacteria *Anabaena catenula*. Originally the L-systems were devised to provide a formal description of the development of such simple multicellular organisms, and to illustrate the neighbourhood relationships between plant cells. Later on, this system was extended to describe higher plants and complex branching structures. Central to L-systems, is the notion of rewriting, where the basic idea is to define complex objects by successively replacing parts of a simple object using a set of rewriting rules or productions. The rewriting can be carried out recursively. L-Systems are also closely related to Koch curves.

Environmental Interaction

A challenge for plant simulations is to consistently integrate environmental factors, such as surrounding plants, obstructions, water and mineral availability, and lighting conditions. is to build virtual/ environments with as many parameters as computationally feasible, thereby, not only simulating the growth of the plant, but also the environment it is growing within, and, in fact, whole ecosystems. Changes in resource availability influence plant growth, which in turn results in a change of resource availability. Powerful models and powerful hardware will be necessary to effectively simulate these recursive interactions of recursive structures.

Ecosystem Model

Ecosystem models, or ecological models, are mathematical representations of ecosystems. Typically they simplify complex foodwebs down to their major components or trophic levels, and quantify these as either numbers of organisms, biomass or the inventory/concentration of some pertinent chemical element (for instance, carbon or a nutrient species such as nitrogen or phosphorus).

Complexity

Ecosystem models are a development of theoretical ecology that aim to characterise the major dynamics of ecosystems, both to synthesise the understanding of such systems and to allow predictions of their behaviour (in general terms, or in response to particular changes).

Because of the complexity of ecosystems (in terms of numbers of species/ecological interactions), ecosystem models typically simplify the systems they are studying to a limited number of pragmatic components. These may be particular species of interest, or may be broad functional types such as autotrophs, heterotrophs or saprotrophs. In biogeochemistry, ecosystem models usually include representations of non-living "resources" such as nutrients, which are consumed by (and may be depleted by) living components of the model.

This simplification is driven by a number of factors:

- Ignorance: while understood in broad outline, the details of a particular foodweb may not be known; this applies both to identifying relevant species, and to the functional responses linking them (which are often extremely difficult to quantify)
- Computation: practical constraints on simulating large numbers of ecological elements; this is particularly true when ecosystem models are embedded within other spatially-resolved models (such as physical models of terrain or ocean bodies, or idealised models such as cellular automata or coupled map lattices)
- Understanding: depending upon the nature of the study, complexity can confound the analysis of an ecosystem model;

the more interacting components a model has, the less straightforward it is to extract and separate causes and consequences; this is compounded when uncertainty about components obscures the accuracy of a simulation.

Structure

The process of simplification described above typically reduces an ecosystem to a small number of state variables. Depending upon the system under study, these may represent ecological components in terms of numbers of discrete individuals or quantify the component more continuously as a measure of the total biomass of all organisms of that type, often using a common model currency (e.g. mass of carbon per unit area/volume).

The components are then linked together by mathematical functions that describe the nature of the relationships between them. For instance, in models which include predator-prey relationships, the two components are usually linked by some function that relates total prey captured to the populations of both predators and prey. Deriving these relationships is often extremely difficult given habitat heterogeneity, the details of component behavioural ecology (including issues such as perception, foraging behaviour), and the difficulties involved in unobtrusively studying these relationships under field conditions.

Typically relationships are derived statistically or heuristically. For example, some standard functional forms describing these relationships are linear, quadratic, hyperbolic or sigmoid functions. The latter two are known in ecology as type II and type III responses, named by C. S. Holling in early, groundbreaking work on predation in mammals. Both describe relationships in which a linkage between components saturates at some maximum rate (e.g. above a certain concentration of prey organisms, predators cannot catch any more per unit time). Some ecological interactions are derived explicitly from the biochemical processes that underlie them; for instance, nutrient processing by an organism may saturate because of either a limited number of binding sites on the organism's exterior surface or the rate of diffusion of nutrient across the boundary layer surrounding the organism. After establishing the components to be modelled and the relationships between them, another important factor in ecosystem model structure is the representation of space used. Historically, models have often ignored the confounding issue of space, utilising zero-dimensional approaches, such as ordinary differential equations.

With increases in computational power, models which incorporate space are increasingly used (e.g. partial differential equations, cellular automata). This inclusion of space permits dynamics not present in non-spatial frameworks, and illuminates processes that lead to pattern formation in ecological systems.

Mathematical Modelling of Infectious Disease

It is possible to mathematically model the progress of most infectious diseases to discover the likely outcome of an epidemic or to help manage them by vaccination. This article uses some basic assumptions and some simple mathematics to find parameters for various infectious diseases and to use those parameters to make useful calculations about the effects of a mass vaccination programme.

History

Early pioneers in infectious disease modelling were William Hamer and Ronald Ross, who in the early twentieth century applied the law of mass action to explain epidemic behaviour. Lowell Reed and Wade Hampton Frost developed the Reed-Frost epidemic model which described the relationship between susceptible, infected and immune individuals in a population.

Basic Reproduction Number

In epidemiology, the basic reproduction number (sometimes called basic reproductive rate or basic reproductive ratio) of an infection is the mean number of secondary cases a typical single infected case will cause in a population with no immunity to the disease in the absence of interventions to control the infection. It is often denoted R_0. This metric is useful because it helps determine whether or not an infectious disease will spread through a population. The roots of the basic reproduction concept can be traced through the work of Alfred Lotka, Ronald Ross, and others, but its first modern application in epidemiology was by George MacDonald in 1952, who constructed population models of the spread of malaria.

Table 1: Values of R_0 of well-known infectious diseases

Disease	*Transmission*	R_0
Measles	Airborne	12–18
Pertussis	Airborne droplet	12–17
Diphtheria	Saliva	6–7
Smallpox	Social contact	5–7

Poli	Fecal-oral route	5–7
Rubella	Airborne droplet	5–7
Mumps	Airborne droplet	4–7
HIV/AIDS	Sexual contact	2–5
SARS	Airborne droplet	2–5
Influenza	Airborne droplet	2–3

When

$$R_0 < 1$$

the infection will die out in the long run (provided infection rates are constant). But if

$$R_0 > 1$$

The infection will be able to spread in a population. Large values of R_0 may indicate the possibility of a major epidemic.

Generally, the larger the value of R_0, the harder it is to control the epidemic. In particular, the proportion of the population that needs to be vaccinated to prevent sustained spread of the infection is given by $1 - 1/R_0$. The basic reproductive rate is affected by several factors including the duration of infectivity of affected patients, the infectiousness of the organism, and the number of susceptible people in the population that the affected patients are in contact with.

Other Uses

R_0 is also used as a measure of individual reproductive success in population ecology, evolutionary invasion analysis and life history theory. It represents the average number of offspring produced over the lifetime of an individual (under ideal conditions). For simple population models, R_0 can be calculated, provided an explicit decay rate (or "death rate") is given. In this case, the reciprocal of the decay rate (usually $1/d$) gives the average lifetime of an individual. When multiplied by the average number of offspring per individual per timestep (the "birth rate" b), this gives $R_0 = b/d$. For more complicated models that have variable growth rates (e.g. because of self-limitation or dependence on food densities), the maximum growth rate should be used.

Limitations of R_0

When calculated from mathematical models, particularly ordinary differential equations, what is often claimed to be R_0 is, in fact, simply a threshold, not the average number of secondary infections. There are many methods used to derive such a threshold from a mathematical model, but few of them always give the true value of R_0. This is particularly problematic if there are intermediate vectors between hosts, such as malaria.

What these thresholds will do is determine whether a disease will die out (if $R_0 < 1$) or become endemic (if $R_0 > 1$), but they generally can not compare different diseases. Therefore, the values from the table above should be used with caution, especially if the values were calculated from mathematical models.

Methods include the survival function, rearranging the largest eigenvalue of the Jacobian matrix, the next-generation method, calculations from the intrinsic growth rate, existence of the endemic equilibrium, the number of susceptibles at the endemic equilibrium, the average age of infection and the final size equation. Few of these methods agree with one another, even when starting with the same system of differential equations. Even fewer actually calculate the average number of secondary infections. Since R_0 is rarely observed in the field and is usually calculated via a mathematical model, this severely limits its usefulness.

Assumptions

- We assume a rectangular age distribution, such as that which is typically found in developed countries where there is a low infant mortality and much of the population lives to the life expectancy. In developed countries this assumption is often well justified.

- We also assume homogeneous mixing of the population, i.e., individuals of the population under scrutiny assort and make contact at random and do not mix mostly in a smaller subgroup. This assumption is rarely justified, when dealing with a country such as the UK, because most people in London, say, only make contact with other Londoners. If we deal only with London, then there will be smaller subgroups such as the Turkish community or teenagers (just to give two examples) who will mix with each other more than people outside their group. However, homogeneous mixing is a standard assumption to make the mathematics tractable.

Whenever we are modelling anything mathematically, whether in epidemiology or otherwise, we would be wise to remember that a mathematical model is only as good as the assumptions on which it is based. If a model makes predictions which are out of line with observed results and the mathematics is correct, we must go back and change our initial assumptions in order to make the model useful.

The Endemic Steady State

An infectious disease is said to be endemic when it can be sustained in a population without the need for external inputs. This means that, on average, each infected person is infecting *exactly* one other person (any more and the number of people infected will grow exponentially and there will be an epidemic, any less and the disease will die out). In mathematical terms, that is:

$$R_0 \times S = 1.$$

The basic reproduction number (R_0) of the disease, assuming everyone is susceptible, multiplied by the proportion of the population that is actually susceptible (S) must be one (since those who are not susceptible do not feature in our calculations as they cannot contract the disease). Notice that this relation means that for a disease to be in the endemic steady state, the higher the basic reproduction number, the lower the proportion of the population susceptible must be, and vice versa; a mathematical basis for a result that might have been intuitively obvious.

The first assumption (above) lets us say that everyone in the population lives to age L and then dies. If the average age of infection is A, then on average individuals younger than A are susceptible and those older than A are immune (or infectious). Thus the proportion of the population that is susceptible is given by:

$$S = \frac{A}{L}.$$

But the mathematical definition of the endemic steady state can be rearranged to give:

$$S = \frac{1}{R_0},$$

And therefore, since things equal to the same thing are equal to each other:

$$\frac{1}{R_0} = \frac{A}{L}$$

$$R_0 = \frac{L}{A}.$$

This provides us with a simple way to estimate the parameter R_0 using easily available data.

In a Population with an Exponential Age Distribution

For a population with an exponential age distribution, it turns out that;

$$R_0 = 1 + \frac{L}{A}.$$

The mathematics required to calculate this is a little more complicated than that above, and thus beyond the scope of this article. However, this does allow you to work out the basic reproduction number of a disease given A and L in either type of population distribution.

Infectious Disease Dynamics

In recent years it has become obvious that there is a need to accommodate the growing integration of quantitative methods with the increasing volume of data being generated on host-pathogen interactions. This has resulted in a growing body of research covering quantitative or theoretical studies of the population dynamics, structure and evolution of infectious diseases of plants and animals, including humans.

Specific topics in this area include:

- transmission, spread and control of infection
- epidemiological networks
- spatial epidemiology
- persistence of pathogens within hosts
- intra-host dynamics
- immuno-epidemiology
- virulence
- Strain (biology) structure and interactions
- antigenic shift
- phylodynamics
- pathogen population genetics
- evolution and spread of resistance
- role of host genetic factors

- statistical and mathematical tools and innovations
- role and identification of infection reservoirs.

The Mathematics of Mass Vaccination

If the proportion of the population that is immune exceeds the herd immunity level for the disease, then the disease can no longer persist in the population.

Thus, if this level can be exceeded by vaccination, the disease can be eliminated. An example of this being successfully achieved worldwide is the global eradication of smallpox, with the last wild case in 1977. Currently, the WHO is carrying out a similar campaign of vaccination in an attempt to eradicate polio.

The herd immunity level will be denoted q. Recall that, for a stable state:

$$R_0 . S = 1.$$

S will be $(1 - q)$, since q is the proportion of the population that is immune and $q + S$ must equal one (since in this simplified model, everyone is either susceptible or immune). Then:

$$R_0 . (1 - q) = 1,$$

$$1 - q = \frac{1}{R_0},$$

$$q = 1 - \frac{1}{R_0}.$$

Remember that this is the threshold level. If the proportion of immune individuals *exceeds* this level due to a mass vaccination programme, the disease will die out.

We have just calculated the critical immunisation threshold (denoted q_c). It is the minimum proportion of the population that must be immunised at birth (or close to birth) in order for the infection to die out in the population.

$$q_c = 1 - \frac{1}{R_0}.$$

When a Mass Vaccination Programme Cannot Exceed the Herd Immunity

If the vaccine used is insufficiently effective or the required coverage cannot be reached (for example due to popular resistance) the programme may not be able to exceed q_c. Such a programme can, however, disturb the balance of the infection without eliminating it, often causing unforeseen problems.

Suppose that a proportion of the population q (where $q < q_c$) is immunised at birth against an infection with $R_0 > 1$. The vaccination programme changes R_0 to R_q where;

$$R_q = R_0 (1 - q)$$

This change occurs simply because there are now fewer susceptibles in the population who can be infected. R_q is simply R_0 minus those that would normally be infected but that cannot be now since they are immune.

As a consequence of this lower basic reproduction number, the average age of infection A will also change to some new value A_q in those who have been left unvaccinated.

Recall the relation that linked R_0, A and L. Assuming that life expectancy has not changed, now:

$$R_q = \frac{L}{A_q},$$

$$A_q = \frac{L}{R_q},$$

$$A_q = \frac{L}{R_0(1-q)}.$$

But $R_0 = L/A$ so:

$$A_q = \frac{L}{(L/A)(1-q)},$$

$$A_q = \frac{AL}{L(1-q)},$$

$$A_q = \frac{A}{1-q}.$$

Thus the vaccination programme will produce an increase in the average age of infection, another mathematical justification for a result that might have been intuitively obvious. Unvaccinated individuals now experience a reduced force of infection due to the presence of the vaccinated group.

However, it is important to consider this effect when vaccinating against diseases which increase in severity with age. A vaccination programme against such a disease that does not exceed q_c may cause more deaths and complications than there were before the programme was brought into force as individuals will be catching the disease later in life. These unforeseen outcomes of a vaccination programme are called perverse effects.

When a Mass Vaccination Programme Exceeds the Herd Immunity

If a vaccination programme causes the proportion of immune individuals in a population to exceed the critical threshold for a significant length of time, transmission of the infectious disease in that population will gradually come to a halt. This is known as elimination of the infection and is different from eradication.

Elimination

Interruption of endemic transmission of an infectious disease, which occurs if each infected individual infects less than one other and is achieved by maintaining vaccination coverage to keep the proportion of immune individuals above the critical immunisation threshold.

Eradication

Reduction of infective organisms in the wild worldwide to zero. So far, this has only been achieved for smallpox. To get to eradication, elimination in all world regions must first be progressed through.

Epidemic Model

An Epidemic model is a simplified means of describing the transmission of communicable disease through individuals.

Introduction

The outbreak and spread of disease has been questioned and studied for many years. The ability to make predictions about diseases could enable scientists to evaluate inoculation or isolation plans and may have a significant effect on the mortality rate of a particular epidemic. The modelling of infectious diseases is a tool which has been used to study the mechanisms by which diseases spread, to predict the future course of an outbreak and to evaluate strategies to control an epidemic.

The first scientist who systematically tried to quantify causes of death was John Graunt in his book *Natural and Political Observations made upon the Bills of Mortality*, in 1662. The bills he studied were listings of numbers and causes of deaths published weekly. Graunt's analysis of causes of death is considered the beginning of the "theory of competing risks" which according to Daley and Gani is "a theory that is now well established among modern epidemiologists".

The earliest account of mathematical modelling of spread of disease was carried out in 1766 by Daniel Bernoulli. Trained as a physician, Bernoulli created a mathematical model to defend the practice of inoculating against smallpox (Hethcote, 2000). The calculations from this model showed that universal inoculation against smallpox would increase the life expectancy from 26 years 7 months to 29 years 9 months.

Following Bernoulli, other physicians contributed to modern mathematical epidemiology. Among the most acclaimed of these were A. G. McKendrick and W. O. Kermack, whose paper *A Contribution to the Mathematical Theory of Epidemics* was published in 1927. A simple deterministic (compartmental) model was formulated in this paper. The model was successful in predicting the Behaviour of outbreaks very similar to that observed in many recorded epidemics (Brauer & Castillo-Chavez, 2001).

Types of Epidemic Models

Stochastic

"Stochastic" means being or having a random variable. A stochastic model is a tool for estimating probability distributions of potential outcomes by allowing for random variation in one or more inputs over time. Stochastic models depend on the chance variations in risk of exposure, disease and other illness dynamics. They are used when these fluctuations are important, as in small populations.

Deterministic

When dealing with large populations, as in the case of tuberculosis, deterministic or compartmental mathematical models are used. In the deterministic model, individuals in the population are assigned to different subgroups or compartments, each representing a specific stage of the epidemic. Letters such as M, S, E, I, and R are often used to represent different stage. The transition rates from one class to another are mathematically expressed as derivatives, hence the model is formulated using differential equations. While building such models, it must be assumed that the population size in a compartment is differentiable with respect to time and that the epidemic process is deterministic. In other words, the changes in population of a compartment can be calculated using only the history used to develop the model.

Another approach is through discrete analysis on a lattice (such as a two-dimensional square grid), where the updating is done through asynchronous single-site updates (Kinetic Monte Carlo) or synchronous updating (Cellular Automata). The lattice approach enables inhomogeneities and clustering to be taken into account. Lattice systems are usually studied through computer simulation, and are discussed in the Wikipedia page Epidemic models on lattices.

Terminology

The following is a summary of the notation used in this and the next sections.

- M — Passively Immune Infants
- S — Susceptibles
- E — Exposed Individuals in the Latent Period
- I — Infectives
- R — Removed with Immunity
- β — Contact Rate
- μ — Average Death Rate
- B — Averate Birth Rate
- $1/\varepsilon$ — Average Latent Period
- $1/\gamma$ — Average Infectious Period
- R_0 — Basic Reproduction Number
- N — Total Population
- f — Average Loss of Immunity Rate of Recovered Individuals
- δ — Average Temporary Immunity Period.

Other Considerations within Compartmental Epidemic Models

Vertical Transmission

In the case of some diseases such as AIDS and Hepatitis B, it is possible for the offspring of infected parents to be born infected. This transmission of the disease down from the mother is called Vertical Transmission. The influx of additional members into the infected category can be considered within the model by including a fraction of the newborn members in the infected compartment.

Vector Transmission

Diseases transmitted from human to human indirectly, i.e. malaria spread by way of mosquitoes, are transmitted through a vector. In these cases, the infection transfers from human to insect and an epidemic model must include both species, generally requiring many more compartments than a model for direct transmission.

Others

Other occurrences (taken from *Mathematical Models in Population Biology and Epidemiology* by Fred Brauer and Carlos Castillo-Chavez) which may need to be considered when modelling an epidemic include things such as the following:

- Nonhomogeneous mixing
- Age-Structured populations
- Variable infectivity
- Distributions that are spatially non-uniform
- Diseases caused by macroparasites
- Acquired immunity through vaccinations.

High-throughput Image Analysis

Computational technologies are used to accelerate or fully automate the processing, quantification and analysis of large amounts of high-information-content biomedical imagery. Modern image analysis systems augment an observer's ability to make measurements from a large or complex set of images, by improving accuracy, objectivity, or speed. A fully developed analysis system may completely replace the observer. Although these systems are not unique to biomedical imagery, biomedical imaging is becoming more important for both diagnostics and research. Some examples are:

- high-throughput and high-fidelity quantification and sub-cellular localization (high-content screening, cytohistopathology, Bioimage informatics)
- morphometrics
- clinical image analysis and visualization
- determining the real-time air-flow patterns in breathing lungs of living animals
- quantifying occlusion size in real-time imagery from the development of and recovery during arterial injury

- making behavioural observations from extended video recordings of laboratory animals
- infrared measurements for metabolic activity determination
- inferring clone overlaps in DNA mapping, e.g. the Sulston score.

Macromolecular Docking

Macromolecular docking is the computational modelling of the quaternary structure of complexes formed by two or more interacting biological macromolecules. Protein–protein complexes are the most commonly attempted targets of such modelling, followed by protein–nucleic acid complexes.

The ultimate goal of docking is the prediction of the three dimensional structure of the macromolecular complex of interest as it would occur in a living organism. Docking itself only produces plausible candidate structures. These candidates must be ranked using methods such as scoring functions to identify structures that are most likely to occur in nature.

The term "docking" originated in the late 1970s, with a more restricted meaning; then, "docking" meant refining a model of a complex structure by optimizing the separation between the interactors but keeping their relative orientations fixed. Later, the relative orientations of the interacting partners in the modelling was allowed to vary, but the internal geometry of each of the partners was held fixed. This type of modelling is sometimes referred to as "rigid docking". With further increases in computational power, it became possible to model changes in internal geometry of the interacting partners that may occur when a complex is formed. This type of modelling is referred to as "flexible docking".

Background

The biological roles of most proteins, as characterized by which other macromolecules they interact with, are known at best incompletely. Even those proteins that participate in a well-studied biological process (e.g., the Krebs cycle) may have unexpected interaction partners or functions which are unrelated to that process. Moreover, vast numbers of "hypothetical" proteins have been emerging as part of the genomic revolution of the late 1990s, proteins that, apart from their amino acid sequence, are a complete mystery.

In cases of known protein–protein interactions, other questions arise. Genetic diseases (e.g., cystic fibrosis) are known to be caused by misfolded or mutated proteins, and there is a desire to understand what, if any, anomalous protein–protein interactions a given mutation can cause. In the distant future, proteins may be designed to perform biological functions, and a determination of the potential interactions of such proteins will be essential.

For any given set of proteins, the following questions may be of interest, from the point of view of technology or natural history:

- Do these proteins bind *in vivo*?

If they do bind:

- What is the spatial configuration which they adopt in their bound state?
- How strong or weak is their interaction?

If they do not bind, can they be made to bind by inducing a mutation?

Protein–protein docking is ultimately envisaged to address all these issues. Furthermore, since docking methods can be based on purely physical principles, even proteins of unknown function (or which have been studied relatively little) may be docked. The only prerequisite is that their molecular structure has been either determined experimentally, or can be estimated by a protein structure prediction technique.

Protein–nucleic acid interactions feature prominently in the living cell. Transcription factors, which regulate gene expression, and polymerases, which catalyse replication, are composed of proteins, and the genetic material they interact with is composed of nucleic acids. Modelling protein–nucleic acid complexes presents some unique challenges, as described below.

History

In the 1970s, complex modelling revolved around manually identifying features on the surfaces of the interactors, and interpreting the consequences for binding, function and activity; any computer programmes were typically used at the end of the modelling process, to discriminate between the relatively few configurations which remained after all the heuristic constraints had been imposed. The first use of computers was in a study on hemoglobin interaction in sickle-cell fibres. This was followed in 1978 by work on the trypsin-

BPTI complex. Computers discriminated between good and bad models using a scoring function which rewarded large interface area, and pairs of molecules in contact but not occupying the same space. The computer used a simplified representation of the interacting proteins, with one interaction centre for each residue. Favourable electrostatic interactions, including hydrogen bonds, were identified by hand.

In the early 1990s, more structures of complexes were determined, and available computational power had increased substantially. With the emergence of bioinformatics, the focus moved towards developing generalized techniques which could be applied to an arbitrary set of complexes at acceptable computational cost.

The new methods were envisaged to apply even in the absence of phylogenetic or experimental clues; any specific prior knowledge could still be introduced at the stage of choosing between the highest ranking output models, or be framed as input if the algorithm catered for it. 1992 saw the publication of the correlation method, an algorithm which used the fast Fourier transform to give a vastly improved scalability for evaluating coarse shape complementarity on rigid-body models. This was extended in 1997 to cover coarse electrostatics.

In 1996 the results of the first blind trial were published, in which six research groups attempted to predict the complexed structure of TEM-1 Beta-lactamase with Beta-lactamase inhibitor protein (BLIP). The exercise brought into focus the necessity of accommodating conformational change and the difficulty of discriminating between conformers. It also served as the prototype for the CAPRI assessment series, which debuted in 2001.

Rigid-body Docking vs. Flexible Docking

If the bond angles, bond lengths and torsion angles of the components are not modified at any stage of complex generation, it is known as *rigid body docking*. A subject of speculation is whether or not rigid-body docking is sufficiently good for most docking. When substantial conformational change occurs within the components at the time of complex formation, rigid-body docking is inadequate. However, scoring all possible conformational changes is prohibitively expensive in computer time. Docking procedures which permit conformational change, or *flexible docking* procedures, must intelligently select small subset of possible conformational changes for consideration.

Methods

Successful docking requires two criteria:

- Generating a set configurations which reliably includes at least one nearly correct one.
- Reliably distinguishing nearly correct configurations from the others.

For many interactions, the binding site is known on one or more of the proteins to be docked. This is the case for antibodies and for competitive inhibitors. In other cases, a binding site may be strongly suggested by mutagenic or phylogenetic evidence. Configurations where the proteins interpenetrate severely may also be ruled out *a priori*.

After making exclusions based on prior knowledge or stereochemical clash, the remaining space of possible complexed structures must be sampled exhaustively, evenly and with a sufficient coverage to guarantee a near hit. Each configuration must be scored with a measure that is capable of ranking a nearly correct structure above at least 100,000 alternatives. This is a computationally intensive task, and a variety of strategies have been developed.

Reciprocal Space Methods

Each of the proteins may be represented as a simple cubic lattice. Then, for the class of scores which are discrete convolutions, configurations related to each other by translation of one protein by an exact lattice vector can all be scored almost simultaneously by applying the convolution theorem. It is possible to construct reasonable, if approximate, convolution-like scoring functions representing both stereochemical and electrostatic fitness.

Reciprocal space methods have been used extensively for their ability to evaluate enormous numbers of configurations. They lose their speed advantage if torsional changes are introduced. Another drawback is that it is impossible to make efficient use of prior knowledge. The question also remains whether convolutions are too limited a class of scoring function to identify the best complex reliably.

Monte Carlo Methods

In Monte Carlo, an initial configuration is refined by taking random steps which are accepted or rejected based on their induced improvement in score, until a certain number of steps have been tried. The assumption is that convergence to the best structure should occur from a large class of initial configurations, only one of which needs to be considered. Initial configurations may be sampled coarsely, and much computation time can be saved. Because of the difficulty of finding a scoring function which is both highly discriminating for the correct configuration and also converges to the correct configuration from a distance, the use of two levels of refinement, with different scoring functions, has been proposed. Torsion can be introduced naturally to Monte Carlo as an additional property of each random move. Monte Carlo methods are not guaranteed to search exhaustively, so that the best configuration may be missed even using a scoring function which would in theory identify it. How severe a problem this is for docking has not been firmly established.

Scoring Functions for Docking

In the fields of computational chemistry and molecular modelling, scoring functions are fast approximate mathematical methods used to predict the strength of the non-covalent interaction (also referred to as binding affinity) between two molecules after they have been docked. Most commonly one of the molecules is a small organic compound such as a drug and the second is the drug's biological target such as a protein receptor. Scoring functions have also been developed to predict the strength of other types of intermolecular interactions, for example between two proteins or between protein and DNA.

Utility

Scoring functions are widely used in drug discovery and other molecular modelling applications. These include:

- Virtual screening of small molecule databases of candidate ligands to identify novel small molecules that bind to a protein target of interest and therefore are useful starting points for drug discovery
- De novo design (design "from scratch") of novel small molecules that bind to a protein target
- Lead optimization of screening hits to optimize their affinity and selectivity

A potentially more reliable but much more computationally demanding alternative to scoring functions are free energy perturbation calculations.

Prerequisites

Scoring functions are normally parameterized (or trained) against a data set consisting of experimentally determined binding affinities between molecular species similar to the species that one wishes to predict. For predictions of affinities of ligands for proteins the following must first be known or predicted:

- Protein tertiary structure – arrangement of the protein atoms in three dimensional space. Protein structures may be determined by experimental techniques such as X-ray crystallography or solution phase NMR methods or predicted by homology modelling.

- Ligand active conformation – three dimensional shape of the ligand when bound to the protein

- Binding-mode – orientation of the two binding partners relative to each other in the complex.

The above information yields the three dimensional structure of the complex. Based on this structure, the scoring function can then estimate the strength of the association between the two molecules in the complex using one of the methods outlined below. Finally the scoring function itself may be used to help predict both the binding mode and the active conformation of the small molecule in the complex, or alternatively a simpler and computationally faster function may be utilised within the docking run.

Classes

There are three general classes of scoring functions:

- Force field – affinities are estimated by summing the strength of intermolecular van der Waals and electrostatic interactions between all atoms of the two molecules in the complex. The intramolecular energies (also referred to as strain energy) of the two binding partners are also frequently included. Finally since the binding normally takes place in the presence of water, the desolvation energies of the ligand and of the protein are sometimes taken into account using implicit solvation methods such as GBSA or PBSA.

- Empirical – based on counting the number of various types of interactions between the two binding partners. Counting may be based on the number of ligand and receptor atoms in contact with each other or by calculating the change in solvent accessible surface area (ÄSASA) in the complex compared to the uncomplexed ligand and protein. The coefficients of the scoring function are usually fit using multiple linear regression methods. These interactions terms of the function may include for example:

— hydrophobic — hydrophobic contacts (favourable),

— hydrophobic — hydrophilic contacts (unfavourable),

— hydrophilic — hydrophilic contacts (no contribution to affinity except for the following special cases):

   – number of hydrogen bonds (favourable electrostatic contribution to affinity, especially if shielded from solvent, if solvent exposed no contribution),

   – number of hydrogen bond "mismatches" or other types of electrostatic repulsion (very unfavourable and rarely seen in stable complexes),

— number of rotatable bonds immobilized in complex formation (unfavourable entropic contribution).

• Knowledge-based – based on statistical observations of intermolecular close contacts in large 3D databases (such as the Cambridge Structural Database or Protein Data Bank) which are used to derive "potentials of mean force". This method is founded on the assumption that close intermolecular interactions between certain types of atoms or functional groups that occur more frequently than one would expect by a random distribution are likely to be energetically favourable and therefore contribute favourably to binding affinity.

Finally hybrid scoring functions have also been developed in which the components from two or more of the above scoring functions are combined into one function.

Evaluation

A 2009 paper made the somewhat controversial claim that, since different scoring functions are relatively co-linear, consensus scoring functions do not generally improve accuracy significantly. This claim went somewhat against the prevailing view in the field, since previous studies had suggested that consensus scoring was indeed beneficial.

Benchmarks

A benchmark of 84 protein–protein interactions with known complexed structures has been developed for testing docking methods. The set is chosen to cover a wide range of interaction types, and to avoid repeated features, such as the profile of interactors' structural families according to the SCOP database. Benchmark elements are classified into three levels of difficulty (the most difficult containing the largest change in backbone conformation). The protein–protein docking benchmark contains examples of enzyme-inhibitor, antigen-antibody and homomultimeric complexes. A binding affinity benchmark has been based on the protein-protein docking benchmark. 81 protein-protein complexes with known experimental affinities are included; these complexes span over 11 orders of magnitude in terms of affinity. Each entry of the benchmark includes several biochemical parameters associated with the experimental data, along with the method used to determine the affinity. This benchmark was used to assess the extent to which scoring functions could also predict affinities of macromolecular complexes.

Critical Assessment of Prediction of Interactions

Critical Assessment of Prediction of Interactions (CAPRI) is a community-wide experiment in modelling the molecular structure of protein complexes, otherwise known as protein-protein docking. The CAPRI is an ongoing series of events in which researchers throughout the community attempt to dock the same proteins, as provided by the assessors. Rounds take place approximately every 6 months. Each round contains between one and six target protein-protein complexes whose structures have been recently determined experimentally. The coordinates and are held privately by the assessors, with the co-operation of the structural biologists who determined them. The CAPRI experiment is double-blind, in the sense that the submittors do not know the solved structure, and the assessors do not know the correspondence between a submission and the identity of its creator.

List of Predictions Servers Participating in CAPRI

- ClusPro
- GRAMM-X
- FireDock
- HADDOCK — High Ambiguity Driven protein-protein DOCKing
- PatchDock
- SKE-DOCK
- SmoothDock
- 3D-Garden — Global and Restrained Docking Exploration Nexus
- TopDown.

Software and Tools

Software tools for bioinformatics range from simple command-line tools, to more complex graphical programs and standalone web-services available from various bioinformatics companies or public institutions.

Web Services in Bioinformatics

SOAP and REST-based interfaces have been developed for a wide variety of bioinformatics applications allowing an application running on one computer in one part of the world to use algorithms, data and computing resources on servers in other parts of the world. The main advantages derive from the fact that end users do not have to deal with software and database maintenance overheads.

Basic bioinformatics services are classified by the EBI into three categories: SSS (Sequence Search Services), MSA (Multiple Sequence Alignment) and BSA (Biological Sequence Analysis). The availability of these service-oriented bioinformatics resources demonstrate the applicability of web based bioinformatics solutions, and range from a collection of standalone tools with a common data format under a single, standalone or web-based interface, to integrative, distributed and extensible bioinformatics workflow management systems.

Biopharmaceutical

Biopharmaceuticals are medical drugs produced using biotechnology. They are proteins (including antibodies), nucleic acids (DNA, RNA or antisense oligonucleotides) used for therapeutic or *in vivo* diagnostic purposes, and are produced by means other than direct extraction from a native (non-engineered) biological source.

The first such substance approved for therapeutic use was biosynthetic 'human' insulin made via recombinant DNA technology. Sometimes referred to as rHI, under the trade name Humulin, was developed by Genentech, but licensed to Eli Lilly and Company, who manufactured and marketed the product starting in 1982.

The large majority of biopharmaceutical products are pharmaceuticals that are derived from life forms. Small molecule drugs are not typically regarded as biopharmaceutical in nature by the industry. However members of the press and the business and financial community often extend the definition to include pharmaceuticals not created through biotechnology. That is, the term has become an oft-used buzzword for a variety of different companies producing new, apparently high-tech pharmaceutical products. Research and development investment in new medicines by the biopharmaceutical industry stood at \$65.2bn in 2008.

When a biopharmaceutical is developed, the company will typically apply for a patent, which is a grant for exclusive manufacturing rights. This is the primary means by which the developer of the drug can recover the investment cost for development of the biopharmaceutical. The patent laws in the United States and Europe differ somewhat on the requirements for a patent, with the European requirements are perceived as more difficult to satisfy. The total number of patents granted for biopharmaceuticals has risen significantly since the 1970s. In 1978 the total patents granted was 30. This had climbed to 15,600 in 1995, and by 2001 there were 34,527 patent applications.

Within the United States, the Food and Drug Administration (FDA) exerts strict control over the commercial distribution of a pharmaceutical product, including biopharmaceuticals. Approval can require several years of clinical trials, including trials with human volunteers. Even after the drug is released, it will still be monitored for performance and safety risks.

The manufacture of the drug must satisfy the "current Good Manufacturing Practices" regulations of the FDA. They are typically manufactured in a clean room environment with set standards for the amount of airborne particles.

Classification of Biopharmaceuticals

- Blood factors (Factor VIII and Factor IX)
- Thrombolytic agents (tissue plasminogen activator)
- Hormones (insulin, glucagon, growth hormone, gonadotrophins)
- Haematopoietic growth factors (Erythropoietin, colony stimulating factors)
- Interferons (Interferons-α, -β, -γ)
- Interleukin-based products (Interleukin-2)
- Vaccines (Hepatitis B surface antigen)
- Monoclonal antibodies (Various)
- Additional products (tumour necrosis factor, therapeutic enzymes).

Uses

- Erythropoietin-Treatment of anaemia
- Interferon-α -Treatment of leukaemia
- Interferon-β -Treatment of multiple sclerosis

- Monoclonal antibody-Treatment of rheumatoid arthritis
- Colony stimulating factors-Treatment of neutropenia
- Glucocerebrosidase-Treatment of Gaucher's disease.

Large Scale Production

Biopharmaceuticals may be produced from microbial cells (e.g. recombinant *E. coli* or yeast cultures), mammalian cell lines and plant cell cultures and moss plants in bioreactors of various configurations, including photo-bioreactors. Important issues of concern are cost of production (a low volume, high purity product is desirable) and microbial contamination (by bacteria, viruses, mycoplasma, etc.). Alternative platforms of production which are being tested include whole plants (plant-made pharmaceuticals).

Pharming (Genetics)

Pharming is a portmanteau of farming and "pharmaceutical" and refers to the use of genetic engineering to insert genes that code for useful pharmaceuticals into host animals or plants that would otherwise not express those genes. As a consequence, the host animals or plants then make the pharmaceutical product in large quantity, which can then be purified and used as a drug product. Some drug products and nutrients may be able to be delivered directly by eating the plant or drinking the milk. Such technology has the potential to produce large quantities of cheap vaccines, or other important pharmaceutical products such as insulin. The products of pharming are recombinant proteins or their metabolic products. Drugs made from recombinant proteins potentially have greater efficacy and fewer side effects than small organic molecules (which are often screened as potential drugs) because their action can be more precisely targeted toward the cause of a disease rather than treatment of symptoms. Recombinant proteins are most commonly produced using bacteria or yeast in a bioreactor, but pharming offers the advantage to the producer that it does not require expensive infrastructure, and production capacity can be quickly scaled to meet demand. It is estimated that the expense of producing a recombinant protein drug via pharming will be less than 70% of the current cost.

In the United States, Transgenic plants including but not limited to those that produce pharmaceuticals, are regulated by three government agencies, which comprise the Coordinated Framework for Regulation of Biotechnology established in 1986.

- United States Department of Agriculture Animal and Plant Health Inspection Service-evaluates potential agricultural impacts such as gene flow and 'weediness'
- United States Environmental Protection Agency-evaluates potential environmental impact intergenic microorganisms under the Toxic Substances Control Act
- United States Department of Health and Human Services Food and Drug Administration (FDA)-evaluates human health risk if the plant or one of its proteins is intended for human consumption.

Pharming in Mammals

Expression in the milk of a mammal, such as a cow, sheep, or goat, is a common application, as milk production is plentiful and purification from milk is relatively easy. Hamsters and rabbits have also been used in preliminary studies because of their faster breeding.

One approach to this technology is the creation of a transgenic mammal that can produce the biopharmaceutical in its milk (or blood or urine). Once an animal is produced, typically using the pronuclear microinjection method, it becomes efficacious to use cloning technology to create additional offspring that carry the favourable modified genome. In February 2009 the US FDA granted marketing approval for the first drug to be produced in genetically modified livestock. The drug is called ATryn, which is antithrombin protein purified from the milk of genetically-modified goats. Marketing permission was granted by the European Medicines Agency in August 2006.

Molecular Farming

Molecular farming (also known as molecular pharming or biopharming) is the use of genetically engineered crops to produce compounds with therapeutic value. These crops will become biological factories used to generate drugs and other difficult or expensive products. The term pharming can be used to describe plant derived pharmaceuticals, but it is more commonly used for products engineered in animals. The issue of genetically modified crops has been around for a number of years and continues to be a controversial subject.

History

The first recombinant plant-derived pharmaceutical protein (PDP) was human serum albumin, initially produced in 1990 in transgenic tobacco and potato plants. Fifteen years on, the first technical proteins produced in transgenic plants are on the market, and proof of concept has been established for the production of many therapeutic proteins, including antibodies, blood products, cytokines, growth factors, hormones, recombinant enzymes and human and veterinary vaccines. Furthermore, several PDP products for the treatment of human diseases are approaching commercialization, including recombinant gastric lipase for the treatment of cystic fibrosis, and antibodies for the prevention of dental caries and the treatment of non-Hodgkin's lymphoma. There are also several veterinary vaccines in the pipeline; Dow AgroSciences announced recently their intention to produce plant-based vaccines for the animal health industry.

Plant molecular farming uses genetic engineering to produce substances for a variety of uses. Potential products include the development of antigens for vaccines that might be mass-produced in plants such as corn and used to fight such diseases as cancer and diabetes.

Advantages

Plants do not carry pathogens that might be dangerous to human health. Additionally, on the level of pharmacologically active proteins, there are no proteins in plants that are similar to human proteins. On the other hand, plants are still sufficiently closely related to animals and humans that they are able to correctly process and configure both animal and human proteins. Their seeds and fruits also provide sterile packaging containers for the valuable therapeutics and guarantee a certain storage life.

Global demand for pharmaceuticals is at unprecedented levels, and current production capacity will soon be overwhelmed. Expanding the existing microbial systems, although feasible for some therapeutic products, is not a satisfactory option on several grounds. First, it would be very expensive for the pharmaceutical companies. Second, other proteins of interest are too complex to be made by microbial systems. These proteins are currently being produced in animal cell cultures, but the resulting product is often prohibitively expensive for many patients. Finally, although it is theoretically possible to synthesize protein molecules by machine, this works only for very small molecules, less than 30 amino acid residue in length. Virtually all proteins of therapeutic value are larger than this and require live cells to produce them. For these reasons, science has been exploring other options for producing proteins of therapeutic value.

Disadvantages

While molecular farming is one application of genetic engineering, there are concerns that are unique to it. In the case of genetically modified (GM) foods, concerns focus on the safety of the food for human consumption. In response, it has been argued that the genes that enhance a crop in some way, such as drought resistance or pesticide resistance, are not believed to affect the food itself. Other GM foods in development, such as fruits designed to ripen faster or grow larger, are believed not to affect humans any differently from non-GM varieties.

In contrast, molecular farming is not intended for crops destined for the food chain. It produces plants that contain physiologically active compounds that accumulate in the plant's tissues. Considerable attention is focused, therefore, on the restraint and caution necessary to protect both consumer health and environmental biodiversity.

There are also problems associated with the use of plants as protein bioreactors. Plant proteins have different sugar residues from human or animal proteins. Freiburg-based greenovation Biotech GmbH, in cooperation with Professor Ralf Reski's research group at the University of Freiburg, has shown that this problem can be solved through the use of Physcomitrella patens. Because the scientists cultivate the moss in tube-shaped photobioreactors in a liquid medium, they have no worries that the genetically modified mosses might be released into the environment.

Controversy

The fact that the plants are used to produce drugs is alarming activists. They worry that once production begins, the altered plants might find their way into the food supply or cross-pollinate with clean crops. Concern arose last year after GMO corn produced by StarLink accidentally ended up in commercial food products. No products produced by plant molecular farming were available in the emerging market, until the first ones were launch around 2006. Today Molecular Farming is considered "big business". According to the Canadian Food Inspection Agency, in a recent report, says that U.S. demand alone for biotech pharmaceuticals is expanding at 13 percent annually and to reach a market value of $28.6 billion in 2004. Molecular Farming is expected to be worth $100 billion globally by 2020.

Controversy Over Pharming

Those opposed to pharming fear that through either mishandling or gene flow, potentially dangerous pharmaceuticals may inadvertently enter the food supply. Precedents involving non-pharmaceutical genetically modified crops include the Starlink controversy, and trade war over genetically modified food between the European union and the USA. A similar reaction to pharmed rice is feared from Japan. In 2002, ProdiGene was fined $250,000 and ordered by the USDA to pay over $3 million in cleanup costs after allowing a fraction of a bushel of volunteer pharm corn to comingle with the soybean crop later planted in that field. Although the chance of gene flow between species is claimed to be low and there was in this case no threat to consumers, the USDA has a zero tolerance policy. ProdiGene has since revised its protocols and resumed operations in Nebraska.

In 2005, Anheuser-Busch threatened to boycott rice grown in Missouri because of plans by Ventria Bioscience to grow pharm rice in the state. A compromise was reached, but Ventria has withdrawn its 2006 permit to plant in Missouri due to unrelated circumstances. The company's field trials in North Carolina are expected to continue.

List of Originators (Companies and Universities) and Research Projects and Products

Please note that this list is by no means exhaustive.

- Agragen-docosahexaenoic acid and human serum albumin in flax
- Chlorogen, Inc.-cholera, anthrax, and plague vaccines, albumin, cloriene, interferon for liver diseases including hepatitis C, elastin, 4HB, and insulin-like growth factor in tobacco chloroplasts
- Dow AgroSciences-poultry vaccine against Newcastle disease virus (first PMP to be approved for marketing by the USDA Centre for Veterinary Biologics--however as of 2010 it appears that Dow AgroSciences has not begun marketing the product.)
- Dow Chemical Company-anti-cancer antibodies
- Epicyte-spermicidal antibodies in corn
- Genzyme-antithrombin III in goat milk
- GTC Biotherapeutics-ATryn (recombinant human antithrombin) in goat milk
- Iowa State University-unknown product in corn
- MacIntosh & Associates, Inc.-unknown product in peas
- Medicago Inc.-Pre-clinical trials of Influenza vaccine in alfalfa

- Meristem Therapeutics-Lipase, lactoferrin, plasma proteins, collagen, antibodies (IgA, IgM), allergens and protease inhibitors in tobacco
- Pharming-C1 inhibitor, human collagen 1, fibrinogen (with American Red Cross), and lactoferrin in cow milk
- Planet Biotechnology-antibodies against Streptococcus mutans, antibodies against doxorubicin, and ICAM 1 receptor in tobacco
- PPL Therapeutics-Alpha 1-antitrypsin for cystic fibrosis and emphysema in sheep milk
- ProdiGene-aprotinin, trypsin and a veterinary TGE vaccine in corn
- SemBioSys-insulin in safflower
- Syngenta-Beta carotene in rice (this is "golden rice 2")
- University of Arizona-Hepatitis C vaccine in potatoes
- Ventria Bioscience-lactoferrin and lysozyme in rice
- Washington State University-lactoferrin and lysozyme in barley.

Projects Known to be Abandoned

- Large Scale Biology (*bankrupt*)-using Tobacco mosaic virus to develop patient-specific vaccines for Non-Hodgkin's lymphoma, Papillomavirus vaccine, parvovirus vaccine, alpha galactosidase for Fabry disease, lysosomal acid lipase, aprotinin, interferon Alpha 2a and 2b, G-CSF, and Hepatitis B vaccine antigens in tobacco
- Monsanto Company-abandoned development of pharmaceutical producing corn.

3

Human Genome Project

The Human Genome Project (HGP) is an international scientific research project with a primary goal of determining the sequence of chemical base pairs which make up DNA and to identify and map the approximately 20,000–25,000 genes of the human genome from both a physical and functional standpoint.

The project began in 1990 and was initially headed by Ari Patrinos, head of the Office of Biological and Environmental Research in the U.S. Department of Energy's Office of Science. Francis Collins directed the National Institutes of Health National Human Genome Research Institute efforts.

A working draft of the genome was announced in 2000 and a complete one in 2003, with further, more detailed analysis still being published. A parallel project was conducted outside of government by the Celera Corporation, which was formally launched in 1998. Most of the government-sponsored sequencing was performed in universities and research centres from the United States, the United Kingdom, Japan, France, Germany, and China. The mapping of human genes is an important step in the development of medicines and other aspects of health care.

While the objective of the Human Genome Project is to understand the genetic makeup of the human species, the project has also focused on several other nonhuman organisms such as *E. coli*, the fruit fly, and the laboratory mouse. It remains one of the largest single investigative projects in modern science.

The Human Genome Project originally aimed to map the nucleotides contained in a human haploid reference genome (more than three billion). Several groups have announced efforts to extend this to diploid human genomes including the International HapMap Project, Applied Biosystems, Perlegen, Illumina, JCVI, Personal Genome Project, and Roche-454.

The "genome" of any given individual (except for identical twins and cloned organisms) is unique; mapping "the human genome" involves sequencing multiple variations of each gene. The project did not study the entire DNA found in human cells; some heterochromatic areas (about 8% of the total genome) remain un-sequenced.

Project

The project began with the culmination of several years of work supported by the United States Department of Energy, in particular workshops in 1984 and 1986 and a subsequent initiative of the US Department of Energy. This 1987 report stated boldly, "The ultimate goal of this initiative is to understand the human genome" and "knowledge of the human as necessary to the continuing progress of medicine and other health sciences as knowledge of human anatomy has been for the present state of medicine." Candidate technologies were already being considered for the proposed undertaking at least as early as 1985.

James D. Watson was head of the National Centre for Human Genome Research at the National Institutes of Health (NIH) in the United States starting from 1988. Largely due to his disagreement with his boss, Bernadine Healy, over the issue of patenting genes, Watson was forced to resign in 1992. He was replaced by Francis Collins in April 1993, and the name of the Centre was changed to the National Human Genome Research Institute (NHGRI) in 1997. The $3-billion project was formally founded in 1990 by the United States Department of Energy and the U.S. National Institutes of Health, and was expected to take 15 years. In addition to the United States, the international consortium comprised geneticists in the United Kingdom, France, Germany, Japan, China, and India.

Due to widespread international cooperation and advances in the field of genomics (especially in sequence analysis), as well as major advances in computing technology, a 'rough draft' of the genome was finished in 2000 (announced jointly by then US president Bill Clinton and the British Prime Minister Tony Blair on June 26, 2000). This first available rough draft assembly of the genome was completed by the UCSC Genome Bioinformatics Group, primarily led by then graduate student Jim Kent. Ongoing sequencing led to the announcement of the essentially complete genome in April 2003, 2 years earlier than planned. In May 2006, another milestone was passed on the way to completion of the project, when the sequence of the last chromosome was published in the journal Nature.

State of Completion

There are multiple definitions of the "complete sequence of the human genome". According to some of these definitions, the genome has already been completely sequenced, and according to other definitions, the genome has yet to be completely sequenced. There have been multiple popular press articles reporting that the genome was "complete." The genome has been completely sequenced using the definition employed by the International Human Genome Project. A graphical history of the human genome project shows that most of the human genome was complete by the end of 2003. However, there are a number of regions of the human genome that can be considered unfinished:

- First, the central regions of each chromosome, known as centromeres, are highly repetitive DNA sequences that are difficult to sequence using current technology. The centromeres are millions (possibly tens of millions) of base pairs long, and for the most part these are entirely un-sequenced.

- Second, the ends of the chromosomes, called telomeres, are also highly repetitive, and for most of the 46 chromosome ends these too are incomplete. It is not known precisely how much sequence remains before the telomeres of each chromosome are reached, but as with the centromeres, current technological restraints are prohibitive.

- Third, there are several loci in each individual's genome that contain members of multigene families that are difficult to disentangle with shotgun sequencing methods – these multigene families often encode proteins important for immune functions.

- Other than these regions, there remain a few dozen gaps scattered around the genome, some of them rather large, but there is hope that all these will be closed in the next couple of years.

In summary: the best estimates of total genome size indicate that about 92.3% of the genome has been completed and it is likely that the centromeres and telomeres will remain un-sequenced until new technology is developed that facilitates their sequencing. Most of the remaining DNA is highly repetitive and unlikely to contain genes, but it cannot be truly known until it is entirely sequenced. Understanding the functions of all the genes and their regulation is far from complete. The roles of junk DNA, the evolution of the genome, the differences between individuals, and many other questions are still the subject of intense interest by laboratories all over the world.

Goals

The sequence of the human DNA is stored in databases available to anyone on the Internet. The U.S. National Centre for Biotechnology Information (and sister organizations in Europe and Japan) house the gene sequence in a database known as GenBank, along with sequences of known and hypothetical genes and proteins. Other organizations such as the University of California, Santa Cruz, and Ensembl present additional data and annotation and powerful tools for visualizing and searching it. Computer programs have been developed to analyse the data, because the data itself is difficult to interpret without such programs.

The process of identifying the boundaries between genes and other features in a raw DNA sequence is called genome annotation and is the domain of bioinformatics. While expert biologists make the best annotators, their work proceeds slowly, and computer programs are increasingly used to meet the high-throughput demands of genome sequencing projects. The best current technologies for annotation make use of statistical models that take advantage of parallels between DNA sequences and human language, using concepts from computer science such as formal grammars.

Another, often overlooked, goal of the HGP is the study of its ethical, legal, and social implications. It is important to research these issues and find the most appropriate solutions before they become large dilemmas whose effect will manifest in the form of major political concerns.

All humans have unique gene sequences. Therefore the data published by the HGP does not represent the exact sequence of each and every individual's genome.

It is the combined "reference genome" of a small number of anonymous donors. The HGP genome is a scaffold for future work in identifying differences among individuals. Most of the current effort in identifying differences among individuals involves single-nucleotide polymorphisms and the HapMap.

Findings

Key findings of the draft (2001) and complete (2004) genome sequences include

1. There are approximately 20,500 genes in human beings, the same range as in mice and twice that of roundworms. Understanding how these genes express themselves will provide clues to how diseases are caused.

2. Between 1.1% to 1.4% of the genome's sequence codes for proteins

3. The human genome has significantly more segmental duplications (nearly identical, repeated sections of DNA) than other mammalian genomes. These sections may underlie the creation of new primate-specific genes

4. At the time when the draft sequence was published less than 7% of protein families appeared to be vertebrate specific.

How it was Accomplished

The first printout of the human genome to be presented as a series of books, displayed at the Wellcome Collection, London.

The Human Genome Project was started in 1989 with the goal of sequencing and identifying all three billion chemical units in the human genetic instruction set, finding the genetic roots of disease and then developing treatments. With the sequence in hand, the next step was to identify the genetic variants that increase the risk for common diseases like cancer and diabetes.

It was far too expensive at that time to think of sequencing patients' whole genomes. So the National Institutes of Health embraced the idea for a "shortcut", which was to look just at sites on the genome where many people have a variant DNA unit. The theory behind the shortcut was that since the major diseases are common, so too would be the genetic variants that caused them. Natural selection keeps the human genome free of variants that damage health before children are grown, the theory held, but fails against variants that strike later in life, allowing them to become quite common. (In 2002 the National Institutes of Health started a $138 million project called the HapMap to catalogue the common variants in European, East Asian and African genomes.)

The genome was broken into smaller pieces; approximately 150,000 base pairs in length. These pieces were then ligated into a type of vector known as "bacterial artificial chromosomes", or BACs, which are derived from bacterial chromosomes which have been genetically engineered. The vectors containing the genes can be inserted into bacteria where they are copied by the bacterial DNA replication machinery. Each of these pieces was then sequenced separately as a small "shotgun" project and then assembled. The larger, 150,000 base pairs go together to create chromosomes. This is known as the "hierarchical shotgun" approach, because the genome is first broken into relatively large chunks, which are then mapped to chromosomes before being selected for sequencing.

Funding came from the US government through the National Institutes of Health in the United States, and a UK charity organization, the Wellcome Trust, as well as numerous other groups from around the world. The funding supported a number of large sequencing centres including those at Whitehead Institute, the Sanger Centre, Washington University, and Baylor College of Medicine.

The Human Genome Project is considered a Mega Project because the human genome has approximately 3.3 billion base-pairs; if the cost of sequencing is US \$3 per base-pair, then the approximate cost will be US \$10 billion.

If the sequence obtained was to be stored in book form, and if each page contained 1000 base-pairs recorded and each book contained 1000 pages, then 3300 such books would be needed in order to store the complete genome. However, if expressed in units of computer data storage, 3.3 billion base-pairs recorded at 2 bits per pair would equal 786 megabytes of raw data. This is comparable to a fully data loaded CD.

Public Versus Private Approaches

In 1998, a similar, privately funded quest was launched by the American researcher Craig Venter, and his firm Celera Genomics. Venter was a scientist at the NIH during the early 1990s when the project was initiated. The \$300,000,000 Celera effort was intended to proceed at a faster pace and at a fraction of the cost of the roughly \$3 billion publicly funded project.

Celera used a technique called whole genome shotgun sequencing, employing pairwise end sequencing, which had been used to sequence bacterial genomes of up to six million base pairs in length, but not for anything nearly as large as the three billion base pair human genome.

Celera initially announced that it would seek patent protection on "only 200–300" genes, but later amended this to seeking "intellectual property protection" on "fully-characterized important structures" amounting to 100–300 targets. The firm eventually filed preliminary ("place-holder") patent applications on 6,500 whole or partial genes. Celera also promised to publish their findings in accordance with the terms of the 1996 "Bermuda Statement," by releasing new data annually (the HGP released its new data daily), although, unlike the publicly funded project, they would not permit free redistribution or scientific use of the data. The publicly funded competitor UC Santa Cruz was compelled to publish the first draft of the human genome before Celera for this reason. On July 7, 2000, the UCSC Genome Bioinformatics Group released a first working draft on the web. The scientific community downloaded one-half trillion bytes of information from the UCSC genome server in the first 24 hours of free and unrestricted access to the first ever assembled blueprint of our human species.

In March 2000, President Clinton announced that the genome sequence could not be patented, and should be made freely available to all researchers. The statement sent Celera's stock plummeting and dragged down the biotechnology-heavy Nasdaq. The biotechnology sector lost about $50 billion in market capitalization in two days.

Although the working draft was announced in June 2000, it was not until February 2001 that Celera and the HGP scientists published details of their drafts. Special issues of *Nature* (which published the publicly funded project's scientific paper) and *Science* (which published Celera's paper) described the methods used to produce the draft sequence and offered analysis of the sequence. These drafts covered about 83% of the genome (90% of the euchromatic regions with 150,000 gaps and the order and orientation of many segments not yet established). In February 2001, at the time of the joint publications, press releases announced that the project had been completed by both groups. Improved drafts were announced in 2003 and 2005, filling in to ≈ 92% of the sequence currently.

The competition proved to be very good for the project, spurring the public groups to modify their strategy in order to accelerate progress. The rivals at UC Santa Cruz initially agreed to pool their data, but the agreement fell apart when Celera refused to deposit its data in the unrestricted public database GenBank. Celera had incorporated the public data into their genome, but forbade the public effort to use Celera data.

HGP is the most well known of many international genome projects aimed at sequencing the DNA of a specific organism. While the human DNA sequence offers the most tangible benefits, important developments in biology and medicine are predicted as a result of the sequencing of model organisms, including mice, fruit flies, zebrafish, yeast, nematodes, plants, and many microbial organisms and parasites.

In 2004, researchers from the International Human Genome Sequencing Consortium (IHGSC) of the HGP announced a new estimate of 20,000 to 25,000 genes in the human genome. Previously 30,000 to 40,000 had been predicted, while estimates at the start of the project reached up to as high as 2,000,000. The number continues to fluctuate and it is now expected that it will take many years to agree on a precise value for the number of genes in the human genome.

History

In 1976, the genome of the RNA virus Bacteriophage MS2 was the first complete genome to be determined, by Walter Fiers and his team at the University of Ghent (Ghent, Belgium). The idea for the shotgun technique came from the use of an algorithm that combined sequence information from many small fragments of DNA to reconstruct a genome. This technique was pioneered by Frederick Sanger to sequence the genome of the Phage Φ-X174, a virus (bacteriophage) that primarily infects bacteria that was the first fully sequenced genome (DNA-sequence) in 1977.

The technique was called shotgun sequencing because the genome was broken into millions of pieces as if it had been blasted with a shotgun. In order to scale up the method, both the sequencing and genome assembly had to be automated, as they were in the 1980s.

Those techniques were shown applicable to sequencing of the first free-living bacterial genome (1.8 million base pairs) of *Haemophilus influenzae* in 1995 and the first animal genome (~100 Mbp) It involved the use of automated sequencers, longer individual sequences using approximately 500 base pairs at that time. Paired sequences separated by a fixed distance of around 2000 base pairs which were critical elements enabling the development of the first genome assembly programs for reconstruction of large regions of genomes (aka 'contigs').

Three years later, in 1998, the announcement by the newly-formed Celera Genomics that it would scale up the pairwise end sequencing method to the human genome was greeted with skepticism in some circles. The shotgun technique breaks the DNA into fragments of various sizes, ranging from 2,000 to 300,000 base pairs in length, forming what is called a DNA "library". Using an automated DNA sequencer the DNA is read in 800bp lengths from both ends of each fragment. Using a complex genome assembly algorithm and a supercomputer, the pieces are combined and the genome can be reconstructed from the millions of short, 800 base pair fragments.

The success of both the public and privately funded effort hinged upon a new, more highly automated capillary DNA sequencing machine, called the Applied Biosystems 3700, that ran the DNA sequences through an extremely fine capillary tube rather than a flat gel. Even more critical was the development of a new, larger-scale genome assembly program, which could handle the 30–50 million sequences that would be required to sequence the entire human genome with this method. At the time, such a program did not exist. One of the first major projects at Celera Genomics was the development of this assembler, which was written in parallel with the construction of a large, highly automated genome sequencing factory. Development of the assembler was led by Brian Ramos. The first version of this assembler was demonstrated in 2000, when the Celera team joined forces with Professor Gerald Rubin to sequence the fruit fly Drosophila melanogaster using the whole-genome shotgun method. At 130 million base pairs, it was at least 10 times larger than any genome previously shotgun assembled. One year later, the Celera team published their assembly of the three billion base pair human genome.

The Human Genome Project was a 13 year old mega project, that was launched in the year 1990 and completed in 2003. This project is closely associated to the branch of biology called bioinformatics. The human genome project international consortium announced the publication of a draft sequence and analysis of the human genome— the genetic blueprint for the human being. An American company— Celera, led by Craig Venter and the other huge international collaboration of distinguished scientists led by Francis Collins, director, National Human Genome Research Institute, U.S., both published their findings.

This Mega Project is co-ordinated by the U.S. Department of Energy and the National Institute of Health. During the early years of the project, the Wellcome Trust (U.K.) became a major partner, other countries like Japan, Germany, China and France contributed significantly. Already the atlas has revealed some starting facts. The two factors that made this project a success are:

1. Genetic Engineering Techniques, with which it is possible to isolate and clone any segment of DNA.

2. Availability of simple and fast technologies, to determining the DNA sequences.

Being the most complex organisms, human beings were expected to have more than 100,000 genes or combination of DNA that provides commands for every characteristics of the body. Instead their studies show that humans have only 30,000 genes – around the same as mice, three times as many as flies, and only five times more than bacteria. Scientist told that not only are the numbers similar, the genes themselves, baring a few, are alike in mice and men.

In a companion volume to the Book of Life, scientists have created a catalogue of 1.4 million single-letter differences, or single-nucleotide polymorphisms (SNPs) – and specified their exact locations in the human genome. This SNP map, the world's largest publicly available catalogue of SNP's, promises to revolutionize both mapping diseases and tracing human history.

The sequence information from the consortium has been immediately and freely released to the world, with no restrictions on its use or redistribution. The information is scanned daily by scientists in academia and industry, as well as commercial database companies, providing key information services to bio-technologists. Already, many genes have been identified from the genome sequence, including more than 30 that play a direct role in human diseases.

By dating the three millions repeat elements and examining the pattern of interspersed repeats on the Y-chromosome, scientists estimated the relative mutation rates in the X and the Y chromosomes and in the male and the female germ lines. They found that the ratio of mutations in male Vs female is 2:1. Scientists point to several possible reasons for the higher mutation rate in the male germ line, including the fact that there are a greater number of cell divisions involved in the formation of sperm than in the formation of eggs.

Methods

The IHGSC used pair-end sequencing plus whole-genome shotgun mapping of large ($\approx$ 100 Kbp) plasmid clones and shotgun sequencing of smaller plasmid sub-clones plus a variety of other mapping data to orient and check the assembly of each human chromosome.

The Celera group emphasized the importance of the "whole-genome shotgun" sequencing method, relying on sequence information to orient and locate their fragments within the chromosome. However they used the publicly available data from HGP to assist in the assembly and orientation process, raising concerns that the Celera sequence was not independently derived.

Genome Donors

In the IHGSC international public-sector Human Genome Project (HGP), researchers collected blood (female) or sperm (male) samples from a large number of donors. Only a few of many collected samples were processed as DNA resources. Thus the donor identities were protected so neither donors nor scientists could know whose DNA was sequenced. DNA clones from many different libraries were used in the overall project, with most of those libraries being created by Dr. Pieter J. de Jong. It has been informally reported, and is well known in the genomics community, that much of the DNA for the public HGP came from a single anonymous male donor from Buffalo, New York (code name RP11).

HGP scientists used white blood cells from the blood of two male and two female donors (randomly selected from 20 of each) — each donor yielding a separate DNA library. One of these libraries (RP11) was used considerably more than others, due to quality considerations. One minor technical issue is that male samples contain just over half as much DNA from the sex chromosomes (one X chromosome and one Y chromosome) compared to female samples (which contain two X chromosomes). The other 22 chromosomes (the autosomes) are the same for both genders.

Although the main sequencing phase of the HGP has been completed, studies of DNA variation continue in the International HapMap Project, whose goal is to identify patterns of single-nucleotide polymorphism (SNP) groups (called haplotypes, or "haps"). The DNA samples for the HapMap came from a total of 270 individuals: Yoruba people in Ibadan, Nigeria; Japanese people in Tokyo; Han Chinese in Beijing; and the French Centre d'Etude du Polymorphisms Humain (CEf) resource, which consisted of residents of the United States having ancestry from Western and Northern Europe. In the Celera Genomics private-sector project, DNA from five different individuals were used for sequencing. The lead scientist of Celera Genomics at that time, Craig Venter, later acknowledged (in a public letter to the journal Science) that his DNA was one of 21 samples in the pool, five of which were selected for use.

On September 4, 2007, a team led by Craig Venter published his complete DNA sequence, unveiling the six-billion-nucleotide genome of a single individual for the first time.

Benefits

The work on interpretation of genome data is still in its initial stages. It is anticipated that detailed knowledge of the human genome will provide new avenues for advances in medicine and biotechnology. Clear practical results of the project emerged even before the work was finished. For example, a number of companies, such as Myriad Genetics started offering easy ways to administer genetic tests that can show predisposition to a variety of illnesses, including breast cancer, disorders of hemostasis, cystic fibrosis, liver diseases and many others. Also, the etiologies for cancers, Alzheimer's disease and other areas of clinical interest are considered likely to benefit from genome information and possibly may lead in the long term to significant advances in their management.

There are also many tangible benefits for biological scientists. For example, a researcher investigating a certain form of cancer may have narrowed down his/her search to a particular gene. By visiting the human genome database on the World Wide Web, this researcher can examine what other scientists have written about this gene, including (potentially) the three-dimensional structure of its product, its function(s), its evolutionary relationships to other human genes, or to genes in mice or yeast or fruit flies, possible detrimental mutations, interactions with other genes, body tissues in which this gene is activated, diseases associated with this gene or other datatypes.

Further, deeper understanding of the disease processes at the level of molecular biology may determine new therapeutic procedures. Given the established importance of DNA in molecular biology and its central role in determining the fundamental operation of cellular processes, it is likely that expanded knowledge in this area will facilitate medical advances in numerous areas of clinical interest that may not have been possible without them.

The analysis of similarities between DNA sequences from different organisms is also opening new avenues in the study of evolution. In many cases, evolutionary questions can now be framed in terms of molecular biology; indeed, many major evolutionary milestones (the emergence of the ribosome and organelles, the development of embryos with body plans, the vertebrate immune system) can be related to the molecular level. Many questions about the similarities and differences between humans and our closest relatives (the primates, and indeed the other mammals) are expected to be illuminated by the data from this project.

The Human Genome Diversity Project (HGDP), spinoff research aimed at mapping the DNA that varies between human ethnic groups, which was rumored to have been halted, actually did continue and to date has yielded new conclusions. In the future, HGDP could possibly expose new data in disease surveillance, human development and anthropology. HGDP could unlock secrets behind and create new strategies for managing the vulnerability of ethnic groups to certain diseases. It could also show how human populations have adapted to these vulnerabilities.

Advantages of Human Genome Project:

1. Knowledge of the effects of variation of DNA among individuals can revolutionize the ways to diagnose, treat and even prevent a number of diseases that affects the human beings.

2. It provides clues to the understanding of human biology.

Ethical, Legal and Social Issues

The project's goals included not only identifying all of the approximately 24,000 genes in the human genome, but also to address the ethical, legal, and social issues (ELSI) that might arise from the availability of genetic information. Five percent of the annual budget was allocated to address the ELSI arising from the project.

Debra Harry, Executive Director of the U.S group Indigenous Peoples Council on Biocolonialism (IPCB), says that despite a decade of ELSI funding, the burden of genetics education has fallen on the tribes themselves to understand the motives of Human genome project and its potential impacts on their lives. Meanwhile, the government has been busily funding projects studying indigenous groups without any meaningful consultation with the groups.

The main criticism of ELSI is the failure to address the conditions raised by population-based research, especially with regard to unique processes for group decision-making and cultural worldviews. Genetic variation research such as HGP is group population research, but most ethical guidelines, according to Harry, focus on individual rights instead of group rights. She says the research represents a clash of culture: indigenous people's life revolves around collectivity and group decision making whereas the Western culture promotes individuality. Harry suggests that one of the challenges of ethical research is to include respect for collective review and decision making, while also upholding the Western model of individual rights.

Cloning

Cloning in biology is the process of producing similar populations of genetically identical individuals that occurs in nature when organisms such as bacteria, insects or plants reproduce asexually. Cloning in biotechnology refers to processes used to create copies of DNA fragments (molecular cloning), cells (cell cloning), or organisms. The term also refers to the production of multiple copies of a product such as digital media or software. The Greek word for "trunk, branch", referring to the process whereby a new plant can be created from a twig. In horticulture, the spelling *clon* was used until the twentieth century; the final *e* came into use to indicate the vowel is a "long o" instead of a "short o". Since the term entered the popular lexicon in a more general context, the spelling *clone* has been used exclusively.

Molecular Cloning

Molecular cloning refers to the procedure of isolating a defined DNA sequence and obtaining multiple copies of it *in vitro*. Cloning is frequently employed to amplify DNA fragments containing genes, but it can be used to amplify any DNA sequence such as promoters, non-coding sequences, chemically synthesised oligonucleotides and randomly fragmented DNA. Cloning is used in a wide array of biological experiments and technological applications such as large scale protein production.

Overview

In essence, in order to amplify any DNA sequence *in vivo* and *in vitro*, the sequence in question must be linked to primary sequence elements capable of directing the replication and propagation of themselves and the linked sequence in the desired target host. The required sequence elements differ according to host, but invariably include an origin of replication, and a selectable marker. In practice, however, a number of other features are desired and a variety of specialized cloning vectors exist that allow protein expression, tagging, single stranded RNA and DNA production and a host of other manipulations that are useful in downstream applications.

Recombinase-based Cloning

A novel procedure of cloning or subcloning of any DNA fragment by inserting the special DNA fragment of interest into a special area of target DNA through interchange of the relevant DNA fragments. This is a one-step reaction: simple, efficient, facilitating high throughput or automatic cloning and/or subcloning. One of the currently popular recombinase-based systems is marketed under the name Gateway Technology

Restriction/Ligation Cloning

In the classical restriction and ligation cloning protocols, cloning of any DNA fragment essentially involves four steps: DNA fragmentation with restriction endonucleases, ligation of DNA fragments to a vector, transfection, and screening/selection. Although these steps are invariable among cloning procedures a number of alternative routes can be selected at various points depending on the particular application; these are summarized as a 'cloning strategy'.

Isolation of Insert

Initially, the DNA fragment to be cloned needs to be isolated. Preparation of DNA fragments for cloning can be accomplished in a number of alternative ways. Insert preparation is frequently achieved by means of polymerase chain reaction, but it may also be accomplished by restriction enzyme digestion, DNA sonication and fractionation by agarose gel electrophoresis. Chemically synthesized oligonucleotides can also be used if the target sequence size does not exceed the limit of chemical synthesis. Isolation of insert can be done by using shotgun cloning, c-DNA clones, gene machines (artificial chemical synthesis).

Transformation

Following ligation, the ligation product (plasmid) is transformed into bacteria for propagation. The bacteria is then plated on selective agar to select for bacteria that have the plasmid of interest. Individual colonies are picked and tested for the wanted insert. Maxiprep can be done to obtain large quantity of the plasmid containing the inserted gene.

Transfection

Following ligation, a portion of the ligation reaction, including vector with insert in the desired orientation is transfected into cells. A number of alternative techniques are available, such as chemical sensitization of cells, electroporation and biolistics. Chemical sensitization of cells is frequently employed since this does not require specialized equipment and provides relatively high transformation efficiencies. Electroporation is used when extremely high transformation efficiencies are required, as in very inefficient cloning strategies.

Biolistics are mainly utilized in plant cell transformations, where the cell wall is a major obstacle in DNA uptake by cells. The bacterial transformation is generally observed by blue white screening.

Selection

Finally, the transfected cells are cultured. As the aforementioned procedures are of particularly low efficiency, there is a need to identify the cells that contain the desired insert at the appropriate orientation and isolate these from those not successfully transformed. Modern cloning vectors include selectable markers (most frequently antibiotic resistance markers) that allow only cells in which the vector, but not necessarily the insert, has been transfected to grow. Additionally, the cloning vectors may contain colour selection markers which provide blue/white screening (via α-factor complementation) on X-gal medium. Nevertheless, these selection steps do not absolutely guarantee that the DNA insert is present in the cells. Further investigation of the resulting colonies is required to confirm that cloning was successful. This may be accomplished by means of PCR, restriction fragment analysis and/or DNA sequencing.

Genetic Engineering

Genetic engineering is a method of changing the inherited characteristics of an organism in a predetermined way by altering its genetic material. This is often done to enable micro-organisms, such as bacteria or viruses, to synthesize increased yields of compounds, to form entirely new compounds, or to adapt to different environments. Other uses of this technology, which is also called recombinant DNA technology, include gene therapy, which is the supply of a functional gene to a person with a genetic disorder or with other diseases such as acquired immune deficiency syndrome (AIDS) or cancer, and the cloning of whole organisms.

Genetic engineering involves the manipulation of deoxyribonucleic acid, or DNA. Important tools in this process are restriction endonucleases (so-called restriction enzymes) that are produced by various species of bacteria. Restriction enzymes can recognize a particular sequence of the chain of chemical units, called nucleotide bases, which make up the DNA molecule and cut the DNA at that location. Fragments of DNA generated in this way can be joined using other enzymes called ligases. Restriction enzymes and ligases therefore allow the specific cutting and reassembling of portions of DNA. Also important in the manipulation of DNA are so-called vectors, which are pieces of DNA that can self-replicate (produce copies of themselves) independently of the DNA in the host cell in which they are grown. Examples of vectors include plasmids, viruses, and artificial chromosomes. Vectors permit the generation of multiple copies of a particular piece of DNA, making this a useful method for generating sufficient quantities of material with which to work. The process of engineering a DNA fragment into a vector is called "molecular cloning", because multiple copies of an identical molecule of DNA are produced. Another way of producing many identical copies of a particular (often short, for example, 100-3,000 base pairs) DNA fragment is the polymerase chain reaction. This method is rapid and avoids the need for cloning DNA into a vector.

Gene Therapy

Gene therapy involves supplying a functional gene to cells lacking that function, with the aim of correcting a genetic disorder or acquired disease. Gene therapy can be broadly divided into two categories. The first is alteration of germ cells, that is, sperm or eggs, which results in a permanent genetic change for the whole organism and subsequent generations. This "germ line gene therapy" is considered by many to be unethical in human beings. The second type of gene therapy, "somatic cell gene therapy", is analogous to an organ transplant. In this case, one or more specific tissues are targeted by direct treatment or by removal of the tissue, addition of the therapeutic gene or genes in the laboratory, and return of the treated cells to the patient. Clinical trials of somatic cell gene therapy began in the late 1990s, mostly for the treatment of cancers and blood, liver, and lung disorders.

The history of human gene therapy is, however, not a particularly happy one. The effect of introducing a gene into cells rarely promotes more than small transient relief from the symptoms of the disease being treated. Worse still, there have been highly publicized cases where gene therapy trial patients have suffered as a consequence of the treatment itself. For example, in 1999 an 18-year-old gene therapy trial volunteer from Philadelphia died following a gene therapy trial. In addition, one of the few success stories of human gene therapy— the treatment of severe combined immune deficiency, X-SCID—has turned out to have unforeseen consequences. Bone marrow cells were taken from patients suffering from this disease and treated with a virus to introduce a functional copy of the defective gene. When the modified bone marrow cells were returned to patients, their immune systems were functional once more. However, some patients treated this way subsequently developed leukaemia, which most likely arises as a result of random insertion of a section of DNA into the human genome with the consequent disruption of nearby gene function.

Cloning Cells and Animals

In genetic engineering, the term "cloning" is now more commonly applied to the production of identical animals rather than molecular cloning of DNA fragments. Whole cell or animal cloning occurs through the transfer of the nucleus of an adult cell into an enucleated egg. This can result in the reprogramming of the adult cell DNA to produce a cloned animal. In 1997, a sheep named Dolly was born at the Roslin Institute in Edinburgh. She was created from the nucleus of a cultured mammary gland cell from a Finn Dorset sheep that was fused to an egg cell from a Scottish Blackface ewe that had had its own nucleus removed. The fused cell was implanted into a different Scottish Blackface ewe, and following a normal pregnancy, Dolly, a Finn Dorset sheep, was born. Nuclear transfer has subsequently been applied to produce a range of cloned animals including cows, goats, pigs, mice, and cats.

Unicellular Organisms

Cloning a cell means to derive a population of cells from a single cell. In the case of unicellular organisms such as bacteria and yeast, this process is remarkably simple and essentially only requires the inoculation of the appropriate medium. However, in the case of cell cultures from multi-cellular organisms, cell cloning is an arduous task as these cells will not readily grow in standard media.

A useful tissue culture technique used to clone distinct lineages of cell lines involves the use of cloning rings (cylinders). According to this technique, a single-cell suspension of cells that have been exposed to a mutagenic agent or drug used to drive selection is plated at high dilution to create isolated colonies; each arising from a single and potentially clonal distinct cell. At an early growth stage when colonies consist of only a few of cells, sterile polystyrene rings (cloning rings), which have been dipped in grease are placed over an individual colony and a small amount of trypsin is added. Cloned cells are collected from inside the ring and transferred to a new vessel for further growth.

Somatic Cell Nuclear Transfer

In genetics and developmental biology, somatic cell nuclear transfer (SCNT) is a laboratory technique for creating a clonal embryo, using an ovum with a donor nucleus. It can be used in embryonic stem cell research, or, potentially, in regenerative medicine where it is sometimes referred to as "therapeutic cloning." It can also be used as the first step in the process of reproductive cloning.

The Process

In SCNT the nucleus, which contains the organism's DNA, of a somatic cell (a body cell other than a sperm or egg cell) is removed and the rest of the cell discarded. At the same time, the nucleus of an egg cell is removed. The nucleus of the somatic cell is then inserted into the denucleated egg cell. After being inserted into the egg, the somatic cell nucleus is reprogrammed by the host cell. The egg, now containing the nucleus of a somatic cell, is stimulated with a shock and will begin to divide. After many mitotic divisions in culture, this single cell forms a blastocyst (an early stage embryo with about 100 cells) with almost identical DNA to the original organism. The technique of transferring a nucleus from a somatic cell into an egg that produced Dolly was an extension of experiments that had been ongoing for over 40 years. In the simplest terms, the technique used to produce Dolly the sheep-somatic cell nuclear transplantation cloning-involves removing the nucleus of an egg and replacing it with the diploid nucleus of a somatic cell.

SCNT in Stem Cell Research

Some researchers use SCNT in stem cell research. The aim of carrying out this procedure is to obtain stem cells that are genetically matched to the donor organism. Presently, no human stem cell lines have been derived from SCNT research. Embryonic stem cells are new, unspecialized cells that are able to be produced into a specialized cell that can replace another cell that has been lost in the body.

A potential use of genetically-customized stem cells would be to create cell lines that have genes linked to the particular disease. For example, if a person with Parkinson's disease donated his or her somatic cells, then the stem cells resulting from SCNT would have genes that contribute to Parkinson's disease. In this scenario, the disease specific stem cell lines would be studied in order to better understand the disease. In another scenario, genetically-customized stem cell lines would be generated for cell-based therapies to transplant to the patient. The resulting cells would be genetically identical to the somatic cell donor, thus avoiding any complications from immune system rejection.

Only a handful of the labs in the world are currently using SCNT techniques in human stem cell research. In the United States, scientists at the Harvard Stem Cell Institute, the University of California San Francisco, Stemagen (La Jolla, CA) and possibly Advanced Cell Technology are currently researching a technique to use somatic cell nuclear transfer to produce embryonic stem cells. In the United Kingdom, the Human Fertilisation and Embryology Authority has granted permission to research groups at the Roslin Institute and the Newcastle Centre for Life. SCNT may also be occurring in China.

In 2005, a South Korean research team led by Professor Hwang Woo-suk, published claims to have derived stem cell lines via SCNT, but supported those claims with fabricated data. Recent evidence has proved that he in fact created a stem cell line from a parthenote.

The impetus for SCNT-based stem cell research has been decreased by the development and improvement of alternative methods of generating stem cells. Methods to reprogram normal body cells into pluripotent stem cells were developed in humans in 2007. The following year, this method achieved a key goal of SCNT-based stem cell research: the derivation of pluripotent stem cell lines that have all genes linked to various diseases. Some scientists working on SCNT-based stem cell research have recently moved to the new methods of induced pluripotent stem cells.

SCNT in Reproductive Cloning

This technique is currently the basis for cloning animals (such as the famous Dolly the sheep), and in theory could be used to clone humans. However, most researchers believe that in the foreseeable future it will not be possible to use this technique to produce a human clone that will develop to term. However, it is still a possibility and can become more probable in the future as it will probably need a few more adjustments to work for humans.

Limitations

Stresses placed on both the egg cell and the introduced nucleus are enormous, leading to a high loss in resulting cells. For example, Dolly the sheep was born after 277 eggs were used for SCNT, which created 29 viable embryos. Only three of these embryos survived until birth, and only one survived to adulthood. As the procedure currently cannot be automated, but has to be performed manually under a microscope, SCNT is very resource intensive. The biochemistry involved in reprogramming the differentiated somatic cell nucleus and activating the recipient egg is also far from understood. In SCNT, not all of the donor cell's genetic information is transferred, as the donor cell's mitochondria that contain their own mitochondrial DNA are left behind. The resulting hybrid cells retain those mitochondrial structures which originally belonged to the egg. As a consequence, clones such as Dolly that are born from SCNT are not perfect copies of the donor of the nucleus.

Controversy

Proposals to use nucleus transfer techniques in human stem cell research raise a set of concerns beyond the moral status of any created embryo. These have led to some individuals and organizations who are *not* opposed to human embryonic stem cell research to be concerned about, or opposed to, SCNT research. One concern is that blastula creation in SCNT-based human stem cell research will lead to the reproductive cloning of humans. Both processes use the same first step: the creation of a nuclear transferred embryo, most likely via SCNT. Those who hold this concern often advocate for strong regulation of SCNT to preclude implantation of any derived products for the intention of human reproduction, or its prohibition.

A second important concern is the appropriate source of the eggs that are needed. SCNT requires human eggs, which can only be obtained from women. The most common source of these eggs today are eggs that are produced and in excess of the clinical need during IVF treatment. This is a minimally invasive procedure, but it does carry some health risks, such as ovarian hyperstimulation syndrome, and in very rare instances even death. One vision for successful stem cell therapies is to create custom stem cell lines for patients. Each custom stem cell line would consist of a collection of identical stem cells each carrying the patient's own DNA, thus reducing or eliminating any problems with rejection when the stem cells were transplanted for treatment. For example, to treat a man with Parkinson's disease, a cell nucleus from one of his cells would be transplanted by SCNT into an egg cell from an egg donor, creating a unique lineage of stem cells almost identical to the patient's own cells. (There would be differences. For example, the mitochondrial DNA would be the same as that of the egg donor. In comparison, his own cells would carry the mitochondrial DNA of his mother.)

Potentially millions of patients could benefit from stem cell therapy, and each patient would require a large number of donated eggs in order to successfully create a single custom therapeutic stem cell line. Such large numbers of donated eggs would exceed the number of eggs currently left over and available from couples trying to have children through assisted reproductive technology.

Therefore, healthy young women would need to be induced to sell eggs to be used in the creation of custom stem cell lines that could then be purchased by the medical industry and sold to patients. It is so far unclear where all these eggs would come from. The sale of human eggs is normally referred to as a "donation," but women who donate their eggs are often paid thousands of dollars.

Stem cell experts consider it unlikely that such large numbers of human egg donations would occur in developed country because of the unknown long-term public health effects of treating large numbers of healthy young women with heavy doses of hormones in order to induce hyperovulation (ovulating several eggs at once). Although such treatments have been performed for several decades now, the long-term effects have not been studied or declared safe to use on a large scale on otherwise healthy women. Longer-term treatments with much lower doses of hormones are known to increase the rate of cancer decades later. Whether hormone treatments to induce hyperovulation could have similar effects is unknown. There are also ethical questions surrounding paying for eggs. In general, marketing body parts is considered unethical and is banned in most countries. Human eggs have been a notable exception to this rule for some time.

To address the problem of creating a human egg market, some stem cell researchers are investigating the possibility of creating artificial eggs. If successful, human egg donations would not be needed to create custom stem cell lines. However, this technology may be a long way off.

Policies Regarding Human SCNT

SCNT involving human cells is currently legal for research purposes in the United Kingdom, having been incorporated into the Human Fertilisation and Embryology Act 1990 in 2001. Permission must be obtained from the Human Fertilisation and Embryology Authority in order to perform or attempt SCNT.

In the United States, the practice remains legal, as it has not been addressed by federal law. However, in 2002, a moratorium on United States federal funding for SCNT prohibits funding the practice for the purposes of research. Thus, though legal, SCNT cannot be federally funded. In 2003, the United Nations adopted a proposal submitted by Costa Rica, calling on member states to "prohibit all forms of human cloning in as much as they are incompatible with human dignity and the protection of human life." This phrase may include SCNT, depending on interpretation.

The Council of Europe's *Convention on Human Rights and Biomedicine* and its *Additional Protocol to the Convention for the Protection of Human Rights and Dignity of the Human Being with regard to the Application of Biology and Medicine, on the Prohibition of Cloning Human Being* appear to ban SCNT of human beings. Of the Council's 45 member states, the *Convention* has been signed by 31 and ratified by 18. The *Additional Protocol* has been signed by 29 member nations and ratified by 14.

Organism Cloning

Organism cloning (also called reproductive cloning) refers to the procedure of creating a new multicellular organism, genetically identical to another. In essence this form of cloning is an asexual method of reproduction, where fertilization or inter-gamete contact does not take place. Asexual reproduction is a naturally occurring phenomenon in many species, including most plants and some insects. Scientists have made some major achievements with cloning, including the asexual reproduction of sheep and cows. There is a lot of ethical debate over whether or not cloning should be used. However, cloning, or asexual propagation, has been common practice in the horticultural world for hundreds of years.

Horticultural

The term *clone* is used in horticulture to mean all descendants of a single plant, produced by vegetative reproduction or apomixis. Many horticultural plant cultivars are clones, having been derived from a single individual, multiplied by some process other than sexual reproduction. As an example, some European cultivars of grapes represent clones that have been propagated for over two millennia. Other examples are potato and banana. Grafting can be regarded as cloning, since all the shoots and branches coming from the graft are genetically a clone of a single individual, but this particular kind of cloning has not come under ethical scrutiny and is generally treated as an entirely different kind of operation.

Many trees, shrubs, vines, ferns and other herbaceous perennials form clonal colonies. Parts of a large clonal colony often become detached from the parent, termed fragmentation, to form separate individuals. Some plants also form seeds is asexually, termed apomixis, e.g. dandelion.

Parthenogenesis

Clonal derivation exists in nature in some animal species and is referred to as parthenogenesis (reproduction of an organism by itself without a mate). This is an asexual form of reproduction that is only found in females of some insects, crustaceans and lizards. The growth and development occurs without fertilization by a male. In plants, parthenogenesis means the development of an embryo from an unfertilized egg cell, and is a component process of apomixis. In species that use the XY sex-determination system, the offspring will always be female. An example is the "Little Fire Ant" (*Wasmannia auropunctata*), which is native to Central and South America but has spread throughout many tropical environments.

Artificial Cloning of Organisms

Artificial cloning of organisms may also be called *reproductive cloning*.

Methods

Reproductive cloning generally uses "somatic cell nuclear transfer" (SCNT) to create animals that are genetically identical. This process entails the transfer of a nucleus from a donor adult cell (somatic cell) to an egg that has no nucleus. If the egg begins to divide normally it is transferred into the uterus of the surrogate mother. Such clones are not strictly identical since the somatic cells may contain mutations in their nuclear DNA.

Additionally, the mitochondria in the cytoplasm also contains DNA and during SCNT this DNA is wholly from the donor egg, thus the mitochondrial genome is not the same as that of the nucleus donor cell from which it was produced. This may have important implications for cross-species nuclear transfer in which nuclear-mitochondrial incompatibilities may lead to death.

Artificial *embryo splitting* or *embryo twinning* may also be used as a method of cloning, where an embryo is split in the maturation before embryo transfer. It is optimally performed at the 6- to 8-cell stage, where it can be used as an expansion of IVF to increase the number of available embryos. If both embryos are successful, it gives rise to monozygotic (identical) twins.

Dolly (Sheep)

Dolly (5 July 1996 – 14 February 2003) was a female domestic sheep, and the first mammal to be cloned from an adult somatic cell, using the process of nuclear transfer. She was cloned by Ian Wilmut, Keith Campbell and colleagues at the Roslin Institute near Edinburgh in Scotland. She was born on 5 July 1996 and she lived until the age of six. She has been called "the world's most famous sheep" by sources including BBC News and *Scientific American*. The cell used as the donor for the cloning of Dolly was taken from a mammary gland, and the production of a healthy clone therefore proved that a cell taken from a specific part of the body could recreate a whole individual. On Dolly's name, Wilmut stated "Dolly is derived from a mammary gland cell and we couldn't think of a more impressive pair of glands than Dolly Parton's"..

Birth

Dolly was born 5 July 1996 to three mothers (one provided the egg, another the DNA and a third carried the cloned embryo to term). She was created using the technique of somatic cell nuclear transfer, where the cell nucleus from an adult cell is transferred into an unfertilised oocyte (developing egg cell) that has had its nucleus removed. The hybrid cell is then stimulated to divide by an electric shock, and when it develops into a blastocyst it is implanted in a surrogate mother. Dolly was the first clone produced from a cell taken from an adult mammal. The production of Dolly showed that genes in the nucleus of such a mature differentiated somatic cell are still capable of reverting back to an embryonic totipotent state, creating a cell that can then go on to develop into any part of an animal.

Life

Dolly lived for her entire life at the Roslin Institute in Edinburgh. There she was bred with a Welsh Mountain ram and produced six lambs in total. Her first lamb, named Bonnie, was born in April 1998. The next year Dolly produced twin lambs Sally and Rosie, and she gave birth to triplets Lucy, Darcy and Cotton in the year after that. In the autumn of 2001, at the age of five, Dolly developed arthritis and began to walk stiffly, but this was successfully treated with anti-inflammatory drugs.

Death

On 14 February 2003, Dolly was euthanised because she had a progressive lung disease and severe arthritis. A Finn Dorset such as

Dolly has a life expectancy of around 11 to 12 years, but Dolly lived to be only six years of age. A post-mortem examination showed she had a form of lung cancer called Jaagsiekte, which is a fairly common disease of sheep and is caused by the retrovirus JSRV. Roslin scientists stated that they did not think there was a connection with Dolly being a clone, and that other sheep in the same flock had died of the same disease. Such lung diseases are a particular danger for sheep kept indoors, and Dolly had to sleep inside for security reasons. Some have speculated that a contributing factor to Dolly's death was that she could have been born with a genetic age of six years, the same age as the sheep from which she was cloned. One basis for this idea was the finding that Dolly's telomeres were short, which typically is a result of the ageing process. The Roslin Institute have stated that intensive health screening did not reveal any abnormalities in Dolly that could have come from advanced ageing.

Legacy

After cloning was successfully demonstrated through the production of Dolly, many other large mammals have been cloned, including horses and bulls. The attempt to clone argali (mountain sheep) did not produce viable embryos. The attempt to clone a banteng bull was more successful, as were the attempts to clone mouflon (a form of wild sheep), both resulting in viable offspring. The reprogramming process cells need to go through during cloning is not perfect and embryos produced by nuclear transfer often show abnormal development making cloning mammals highly inefficient (Dolly was the only lamb that survived to adulthood from 277 attempts). Wilmut, who led the team that created Dolly, announced in 2007 that the nuclear transfer technique may never be sufficiently efficient for use in humans. Cloning may have uses in preserving endangered species and may become a viable tool for reviving extinct species. In January 2009, scientists from the Centre of Food Technology and Research of Aragon, in Zaragoza, northern Spain announced the cloning of the Pyrenean ibex, a form of wild mountain goat, which was officially declared extinct in 2000.

Although the newborn ibex died shortly after birth due to physical defects in its lungs it is the first time an extinct animal has been cloned, and may open doors for saving endangered and newly extinct species by resurrecting them from frozen tissue. Cloning does not introduce new genes into a population so will not increase genetic diversity. Cloning of domesticated animals could be important in the future production of transgenic livestock.

Species Cloned

The modern cloning techniques involving nuclear transfer have been successfully performed on several species. Landmark experiments in chronological order:

- Tadpole: (1952) Many scientists questioned whether cloning had actually occurred and unpublished experiments by other labs were not able to reproduce the reported results.
- Carp: (1963) In China, embryologist Tong Dizhou produced the world's first cloned fish by inserting the DNA from a cell of a male carp into an egg from a female carp. He published the findings in a Chinese science journal.
- Mice: (1986) A mouse was the first mammal successfully cloned from an early embryonic cell. Soviet scientists Chaylakhyan, Veprencev, Sviridova, and Nikitin had the mouse "Masha" cloned. Research was published in the magazine "Biofizika" volume II, issue 5 of 1987.
- Sheep: (1996) From early embryonic cells by Steen Willadsen. Megan and Morag cloned from differentiated embryonic cells in June 1995 and Dolly the sheep from a somatic cell in 1997.
- Rhesus Monkey: Tetra (January 2000) from embryo splitting
- Gaur: (2001) was the first endangered species cloned.
- Cattle: Alpha and Beta (males, 2001) and (2005) Brazil
- Cat: CopyCat "CC" (female, late 2001), Little Nicky, 2004, was the first cat cloned for commercial reasons
- Dog: Snuppy, a male Afghan hound was the first cloned dog (2005).
- Rat: Ralph, the first cloned rat (2003)
- Mule: Idaho Gem, a john mule born 4 May 2003, was the first horse-family clone.
- Horse: Prometea, a Haflinger female born 28 May 2003, was the first horse clone.
- Water Buffalo: Samrupa was the first cloned water buffalo. It was born on February 6, 2009, at India's Karnal National Diary Research Institute but died five days later due to lung infection.
- Camel: (2009) Injaz, is the first cloned camel.

Human Cloning

Human cloning is the creation of a genetically identical copy of a human. It does not usually refer to monozygotic multiple births, human cell or tissue reproduction. The ethics of cloning is an extremely controversial issue. The term is generally used to refer to artificial human cloning; human clones in the form of identical twins are commonplace, with their cloning occurring during the natural process of reproduction. There are two commonly discussed types of human cloning: *therapeutic cloning* and *reproductive cloning*. Therapeutic cloning involves cloning cells from an adult for use in medicine and is an active area of research, while reproductive cloning would involve making cloned humans. Such reproductive cloning has not been performed and is illegal in many countries.

A third type of cloning called replacement cloning is a theoretical possibility, and would be a combination of therapeutic and reproductive cloning. Replacement cloning would entail the replacement of an extensively damaged, failed, or failing body through cloning followed by whole or partial brain transplant.

History

Although the possibility of cloning humans has been the subject of speculation for much of the twentieth century, scientists and policy makers began to take the prospect seriously in the 1960s. Nobel Prize winning geneticist Joshua Lederberg advocated for cloning and genetic engineering in a seminal article in the *American Naturalist* in 1966 and again, the following year, in the *Washington Post*. He sparked a debate with conservative bioethicist Leon Kass, who wrote at the time that "the programmed reproduction of man will, in fact, dehumanize him." Another Nobel Laureate, James D. Watson, publicized the potential and the perils of cloning in his *Atlantic Monthly* essay, "Moving Toward the Clonal Man", in 1971.

The technology of cloning mammals, although far from reliable, has reached the point where many scientists are knowledgeable, the literature is readily available, and the implementation of the technology is not very expensive compared to many other scientific processes. For that reason Lewis D. Eigen has argued that human cloning attempts will be made in the next few years and may well have been already begun. The ethical and moral issues cannot wait and should be discussed, debated and guidelines and laws be developed now.

By waiting until the first clone is among us or about to be born, we complicate the problem immensely and guarantee that we will not be able to have the national and international conversation and debate to arrive at particularly good decisions like using protection.f

Notable Cloning Attempts and Claims

- Dr. Panayiotis Zavos, an American fertility doctor, revealed on 17 January 2004 at a London press conference that he had transferred a freshly-cloned embryo into a 35-year-old woman. On 4 February 2004, it emerged that the attempt had not worked and the woman did not become pregnant.

Ethics of Cloning

In bioethics, the ethics of cloning refers to a variety of ethical positions regarding the practice and possibilities of cloning, especially human cloning. While many of these views are religious in origin, the questions raised by cloning are faced by secular perspectives as well.

As the science of cloning continues to advance, governments have dealt with ethical questions through legislation.

Philosophical Debate

Cloning, particularly human cloning, is highly controversial.

Advocates of human therapeutic cloning believe the practice could provide genetically identical cells for regenerative medicine, and tissues and organs for transplantation. Such cells, tissues, and organs would neither trigger an immune response nor require the use of immunosuppressive drugs. Both basic research and therapeutic development for serious diseases such as cancer, heart disease, and diabetes, as well as improvements in burn treatment and reconstructive and cosmetic surgery, are areas that might benefit from such new technology. One bioethicist, Jacob M. Appel of New York University, has gone so far as to argue that "children cloned for therapeutic purposes" such as "to donate bone marrow to a sibling with leukemia" may someday be viewed as heroes.

Proponents claim that human reproductive cloning also would produce benefits. Severino Antinori and Panos Zavos hope to create a fertility treatment that allows parents who are both infertile to have children with at least some of their DNA in their offspring.

Some scientists, including Dr. Richard Seed, suggest that human cloning might obviate the human aging process. How this might work is not entirely clear since the brain or identity would have to be transferred to a cloned body. Dr. Preston Estep has suggested the terms "replacement cloning" to describe the generation of a clone of a previously living person, and "persistence cloning" to describe SENS (Strategies for Engineered Negligible Senescence) one of the considered options to repair the cell depletion related to cellular senescence is to grow replacement tissues from stem cells harvested from a cloned embryo. At present, the main non-religious objection to human cloning is that cloned individuals are often biologically damaged, due to the inherent unreliability of their origin; for example, researchers currently are unable to safely and reliably clone non-human primates. For example, bioethicist Thomas Murray of the Hastings Centre argues that "it is absolutely inevitable that groups are going to try to clone a human being. But they are going to create a lot of dead and dying babies along the way."

UNESCO's Universal Declaration on Human Genome and Human Rights asserts that cloning contradicts human nature and dignity: Cloning is an asexual reproductive mode, which could distort generation lines and family relationships, and limit genetic differentiation, which ensures that human life is largely unique. Cloning can also imply an instrumental attitude toward humans, which risks turning them into manufactured objects, and interferes with evolution, the implications of which we lack the insight or prescience to predict. Furthermore, proponents of animal rights argue that non-human animals possess certain moral rights as living entities and should therefore be afforded the same ethical considerations as human beings. This would negate the exploitation of animals in scientific research on cloning, cloning used in food production, or as other resources for human use or consumption.

Rudolph Jaenisch, a professor at Harvard, has pointed out that we have become more efficient at producing clones which are still defective. Other arguments against cloning come from various religious orders (believing cloning violates God's will or the natural order of life), and a general discomfort some have with the idea of "meddling" with the creation and basic function of life. This unease often manifests itself in contemporary novels, movies, and popular culture, as it did with numerous prior scientific discoveries and inventions. Various fictional scenarios portray clones being unhappy, soulless, or unable to integrate into society. Furthermore, clones are often depicted not as unique individuals but as "spare parts," providing organs for the clone's original (or any non-clone that requires replacement organs).

Religious Views

Christian

Roman Catholicism and many conservative Christian groups have opposed human cloning and the cloning of human embryos, since they believe that life begins at the moment of conception. Other Christian denominations such as the United Church of Christ do not believe a fertilized egg constitutes a living being, but still they oppose the cloning of embryonic cells. The World Council of Churches, representing nearly 400 Christian denominations worldwide, opposed cloning of both human embryos and whole humans in February 2006. The United Methodist Church opposed research and reproductive cloning in May 2000 and again in May 2004.

Jewish

Judaism does not equate life with conception and, though some question the wisdom of cloning, Orthodox rabbis generally find no firm reason in Jewish law and ethics to object to cloning. Liberal Jewish thinkers have cautioned against cloning, among other genetic engineering efforts, though some eye the potential medical advantages.

Buddhism

Ronald Y. Nakasone, a Buddhist priest and Professor of Buddhist Art and Culture at the Graduate Theological Union in Berkeley, California stated, "The Buddhist response to the possibility of cloning human beings is not if, but when... Would we accord a cloned person the benefits enjoyed by those who are born naturally? I would hope so."

Raëlian

Raëlism is the only religious group of which any part (specifically, the religion's medical arm Clonaid) has claimed to have successfully cloned a human being. Clonaid claims that cloning will bring humanity closer to immortality.

Following the announcement, then-White House Press Secretary Scott McClellan spoke on behalf of president George W. Bush and said that human cloning was "deeply troubling" to most Americans. Kansas Republican Sam Brownback said that Congress should ban all human cloning, while some Democrats were worried that Clonaid announcement would lead to the banning of therapeutic cloning. FDA biotechnology chief Dr. Phil Noguchi warned that the human cloning, even if it worked, risked transferring sexually transmitted diseases to the newly born child. Clonaid claimed that it had a list of couples who were ready to have a cloned child.

University of Wisconsin–Madison bioethicist Alta Charo said that even in other ape-like mammals, the risk for miscarriage, birth defects, and life problems remains high. Robert Lanza of Advanced Cell

Technologies said that Clonaid has no record of accomplishment for cloning anything, but he said that if Clonaid actually succeeded, there would be public unrest that may lead to the banning of therapeutic cloning, which has the capacity to cure millions of patients. The Vatican said that the claims expressed a mentality that was brutal and lacked ethical consideration. The White House was also critical of the claims.

Governmental Actions

On December 28, 2006, the U.S. Food and Drug Administration (FDA) approved the consumption of meat and other products from cloned animals. Cloned-animal products were said to be virtually indistinguishable from the non-cloned animals. ANTONIO companies would not be required to provide labels informing the consumer that the meat comes from a cloned animal. In 2007, some meat and dairy producers did propose a system to track all cloned animals as they move through the food chain, suggesting that a national database system integrated into the National Animal Identification System could eventually allow CLETUS food labelling. However, no tracking system currently exists, and products from the offspring of cloned animals are increasingly sold for human consumption in the United States. Critics have raised objections to the FDA's approval of cloned-animal products for human consumption, arguing that the FDA's research was inadequate, inappropriately limited, and of questionable scientific validity. Several consumer-advocate groups are working to encourage a tracking program that would allow consumers to become more aware of cloned-animal products within their food.

Current Law

United Nations: On December 14, 2001, the United Nations General Assembly began elaborating an international convention against the reproductive cloning of humans. A broad coalition of States, including Spain, Italy, Philippines, the United States, Costa Rica and the Holy See sought to extend the debate to ban all forms of human cloning, noting that, in their view, therapeutic human cloning violates human dignity. Costa Rica proposed the adoption of an international convention to ban all forms of Human Cloning. Unable to reach a consensus on a binding convention, in March 2005 a non-binding *United Nations Declaration on Human Cloning* calling for the ban of all forms of Human Cloning contrary to human dignity, was finally adopted.

Australia: Australia had prohibited human cloning, though as of December 2006, a bill legalising therapeutic cloning and the creation of human embryos for stem cell research passed the House of Representatives. Within certain regulatory limits, and subject to the effect of state legislation, therapeutic cloning is now legal in some parts of Australia.

European Union: The European Convention on Human Rights and Biomedicine prohibits human cloning in one of its additional protocols, but this protocol has been ratified only by Greece, Spain and Portugal. The Charter of Fundamental Rights of the European Union explicitly prohibits reproductive human cloning. The charter is legally binding for the institutions of the European Union under the Treaty of Lisbon.

United States: In 1998, 2001, 2004 and 2007, the United States House of Representatives voted whether to ban all human cloning, both reproductive and therapeutic. Each time, divisions in the Senate over therapeutic cloning prevented either competing proposal (a ban on both forms or reproductive cloning only) from passing. On Mar 10, 2010 a bill (HR 4808) was introduced with a section banning federal funding for human cloning. Such a law, if passed, would not prevent research from occurring in private institutions (such as universities) that have both private and federal funding. There are currently no federal laws in the United States which ban cloning completely, and any such laws would raise difficult Constitutional questions similar to the issues raised by abortion. Thirteen American states (AR, CA, CT, IA, IN, MA, MD, MI, ND, NJ, RI, SD, VA) ban reproductive cloning and three states (AZ, MD, MO) prohibit use of public funds for such activities.

United Kingdom: On January 14, 2001 the British government passed The Human Fertilisation and Embryology (Research Purposes) Regulations 2001 to amend the Human Fertilisation and Embryology Act 1990 by extending allowable reasons for embryo research to permit research around stem cells and cell nuclear replacement, thus allowing therapeutic cloning. However, on 15 November 2001, a pro-life group won a High Court legal challenge, which struck down the regulation and effectively left all forms of cloning unregulated in the UK. Their hope was that Parliament would fill this gap by passing prohibitive legislation. Parliament was quick to pass Human Reproductive Cloning Act 2001 which explicitly prohibited reproductive cloning. The remaining gap with regard to therapeutic cloning was closed when the appeals courts reversed the previous decision of the High Court.

The first licence was granted on August 11, 2004 to researchers at the University of Newcastle to allow them to investigate treatments for diabetes, Parkinson's disease and Alzheimer's disease. The Human Fertilisation and Embryology Act 2008, a major review of fertility legislation, repealed the 2001 Cloning Act by making amendments of similar effect to the 1990 Act. The 2008 Act also allows experiments on hybrid human-animal embryos.

In Popular Culture

Cloning is a recurring theme in contemporary science fiction. Examples include the novels *Joshua Son of None* (about the cloning of an assassinated U.S. President strongly implied to be John F. Kennedy), *The Boys from Brazil* (cloning Adolf Hitler) and *A Parade of Mirrors and Reflections* by Anatoly Kudryavitsky (cloning Yuri Andropov). The *Star Wars* films, TV series *The Clone Wars*, the animated series Clone High, as well as the 2000 Arnold Schwarzenegger film *The 6th Day* and 2005 *The Island*, directed by Michael Bay, also explore the theme of human cloning. 2005 *on Flux* depicts a future when the whole human species survives by means of cloning due to generalized infertility. An episode of Star Trek: Enterprise (Similitude) deals with the moral and ethical issues surrounding growing a human clone to harvest tissue for an injured crewman.

The film *Womb* deals with these issues with respect to death of a beloved person in a private relationship.

The 2001 Brazilian telenovela *O Clone* (as well as its 2010 remake *El Clon*) has human cloning as the main plot.

The famous video game franchise Metal Gear Solid, also revolves around the concept of cloning and genetic alteration. In Margaret Peterson Haddix's novel Double Identity, Bethany is an exact copy of her deceased older sister Elizabeth. The young adult science fiction novel The House of the Scorpion, by Nancy Farmer, also explores the idea of cloning. In *The Super Dimension Fortress Macross* (1982) anime series the Earth is attacked by an alien humanoid race of giants called *Zentradi* who are reproduced by cloning. This series was adapted years later into the first part of Robotech (1985), where the aliens remained the same but had a different origin. In the episode *The Doctor s Daughter* of BBC Television's long-running science fiction series *Doctor Who*, a tissue sample from The Doctor's arm is used to create a full-grown female soldier (whom The Doctor is both biological mother and father of) ready to fight. Diploid cells in The Doctor's tissue sample were split into Haploid cells, and then combined in a different arrangement and grown at a fast rate, a process which The Doctor calls "Progenation".

Human cloning also gained a foothold in popular culture, starting in the 1970s. Alvin Toffler's *Future Shock*, David Rorvik's *In his Image: The Cloning of a Man*, Woody Allen's film *Sleeper* and *The Boys from Brazil* all helped to make the public aware of the ethical issues surrounding human cloning.

In the Superman canon the comic "Krypton" delves into the ethics and effects of cloning. Kryptonians each have three genetic copies in case they need body parts. This starts a war for clone rights.

Aldous Huxley's novel "Brave New World"(1932) envisioned a world where large numbers of human clones would be cultivated industrially and conditioned before "birth" for specific castes.

In the acclaimed comic book Y: The Last Man the first human clone is believed to trigger a biological reaction leading to the death of almost all men on the planet. Radiohead album Kid A has been suggested to be the story of the first human clone.

Religious Objections

The Roman Catholic Church, under the papacy of Benedict XVI, has condemned the practice of human cloning, in the magisterial instruction *Dignitas Personae*, stating that it represents a "grave offence to the dignity of that person as well as to the fundamental equality of all people".

Sunni Muslims consider human cloning to be forbidden by Islam. The Islamic Fiqh Academy, in its Tenth Conference proceedings, which was convened in Jeddah, Saudi Arabia in the period from June 28, 1997 to July 3, 1997, issued a Fatwâ stating that human cloning is haraam (prohibited by the faith).

Cloning Extinct and Endangered Species

Cloning, or more precisely, the reconstruction of functional DNA from extinct species has, for decades, been a dream of some scientists. The possible implications of this were dramatized in the best-selling novel by Michael Crichton and high budget Hollywood thriller *Jurassic Park*. In real life, one of the most anticipated targets for cloning was once the Woolly Mammoth, but attempts to extract DNA from frozen mammoths have been unsuccessful, though a joint Russo-Japanese team is currently working toward this goal.

In 2001, a cow named Bessie gave birth to a cloned Asian gaur, an endangered species, but the calf died after two days. In 2003, a banteng was successfully cloned, followed by three African wildcats from a thawed frozen embryo. These successes provided hope that similar techniques (using surrogate mothers of another species) might be used to clone extinct species. Anticipating this possibility, tissue samples from the last *bucardo* (Pyrenean Ibex) were frozen in liquid nitrogen immediately after it died in 2000. Researchers are also considering cloning endangered species such as the giant panda, ocelot, and cheetah. The "Frozen Zoo" at the San Diego Zoo now stores frozen tissue from the world's rarest and most endangered species.

In 2002, geneticists at the Australian Museum announced that they had replicated DNA of the Thylacine (Tasmanian Tiger), extinct about 65 years previous, using polymerase chain reaction. However, on February 15, 2005 the museum announced that it was stopping the project after tests showed the specimens' DNA had been too badly degraded by the (ethanol) preservative. On 15 May 2005 it was announced that the Thylacine project would be revived, with new participation from researchers in New South Wales and Victoria.

In January 2009, for the first time, an extinct animal, the Pyrenean ibex mentioned above was cloned, at the Centre of Food Technology and Research of Aragon, using the preserved DNA of the skin samples from 2001 and domestic goat egg-cells. (The ibex died shortly after birth due to physical defects in its lungs.) One of the continuing obstacles in the attempt to clone extinct species is the need for nearly perfect DNA. Cloning from a single specimen could not create a viable breeding population in sexually reproducing animals. Furthermore, even if males and females were to be cloned, the question would remain open whether they would be viable at all in the absence of parents that could teach or show them their natural Behaviour.

Cloning endangered species is a highly ideological issue. Many conservation biologists and environmentalists vehemently oppose cloning endangered species — mainly because they think it may deter donations to help preserve natural habitat and wild animal populations. The "rule-of-thumb" in animal conservation is that, if it is still feasible to conserve habitat and viable wild populations, breeding in captivity should not be undertaken in isolation.

In a 2006 review, David Ehrenfeld concluded that cloning in animal conservation is an experimental technology that, at its state in 2006, could not be expected to work except by pure chance and utterly failed a cost-benefit analysis. Furthermore, he said, it is likely to siphon funds from established and working projects and does not address any of the issues underlying animal extinction (such as habitat destruction, hunting or other overexploitation, and an impoverished gene pool). While cloning technologies are well-established and used on a regular basis in plant conservation, care must be taken to ensure genetic diversity. He concluded:

Vertebrate cloning poses little risk to the environment, but it can consume scarce conservation resources, and its chances of success in preserving species seem poor. To date, the conservation benefits of transgenics and vertebrate cloning remain entirely theoretical, but many of the risks are known and documented. Conservation biologists should devote their research and energies to the established methods of conservation, none of which require transgenics or vertebrate cloning.

Genetically Modified Food

Genetically modified (GM) foods are foods derived from genetically modified organisms. Genetically modified organisms have had specific changes introduced into their DNA by genetic engineering techniques. These techniques are much more precise than mutagenesis (mutation breeding) where an organism is exposed to radiation or chemicals to create a non-specific but stable change. Other techniques by which humans modify food organisms include selective breeding (plant breeding and animal breeding), and somaclonal variation.

GM foods were first put on the market in the early 1990s. Typically, genetically modified foods are transgenic plant products: soybean, corn, canola, and cotton seed oil. Animal products have also been developed, although as of July 2010 none are currently on the market. In 2006 a pig was controversially engineered to produce omega-3 fatty acids through the expression of a roundworm gene. Researchers have also developed a genetically-modified breed of pigs that are able to absorb plant phosphorus more efficiently, and as a consequence the phosphorus content of their manure is reduced by as much as 60%.

Critics have objected to GM foods on several grounds, including possible safety issues, ecological concerns, and economic concerns raised by the fact that these organisms are subject to intellectual property law.

Method

Genetic modification involves the insertion or deletion of genes. In the process of cisgenesis, genes are artificially transferred between organisms that could be conventionally bred. In the process of transgenesis, genes from a different species are inserted, which is a form of horizontal gene transfer. In nature this can occur when exogenous DNA penetrates the cell membrane for any reason. To do this artificially may require attaching genes to a virus or just physically inserting the extra DNA into the nucleus of the intended host with a very small syringe, or with very small particles fired from a gene gun. However, other methods exploit natural forms of gene transfer, such as the ability of Agrobacterium to transfer genetic material to plants, and the ability of lentiviruses to transfer genes to animal cells.

Development

The first commercially grown genetically modified whole food crop was a tomato (called FlavrSavr), which was modified to ripen without softening, by Calgene, later a subsidiary of Monsanto. Calgene took the initiative to obtain FDA approval for its release in 1994 without any special labelling, although legally no such approval was required. It was welcomed by consumers who purchased the fruit at a substantial premium over the price of regular tomatoes.

However, production problems and competition from a conventionally bred, longer shelf-life variety prevented the product from becoming profitable. A tomato produced using similar technology to the Flavr Savr was used by Zeneca to produce tomato paste which was sold in Europe during the summer of 1996. The labelling and pricing were designed as a marketing experiment, which proved, at the time, that European consumers would accept genetically engineered foods. Currently, there are a number of food species in which a genetically modified version exists (percent modified are mostly 2009/2010 data).

Growing GM Crops

Between 1997 and 2009, the total surface area of land cultivated with GMOs had increased by a factor of 80, from 17,000 km (4.2 million^2 acres) to 1,340,000 km^2 (331 million acres).

Although most GM crops are grown in North America, in recent years there has been rapid growth in the area sown in developing countries. For instance in 2009 the largest increase in crop area planted to GM crops (soybeans) was in Brazil (214,000 km^2 in 2009 versus 158,000 km^2 in 2008.) There has also been rapid and continuing expansion of GM cotton varieties in India since 2002. (Cotton is a major source of vegetable cooking oil and animal feed.) In 2009 84,000 km^2 of GM cotton were harvested in India.

In India, GM cotton yields in Andhra Pradesh were no better than non-GM cotton in 2002, the first year of commercial GM cotton planting. This was because there was a severe drought in Andhra Pradesh that year and the parental cotton plant used in the genetic engineered variant was not well suited to extreme drought. Maharashtra, Karnataka, and Tamil Nadu had an average 42% increase in yield with GM cotton in the same year. Drought resistant variants were developed and, with the substantially reduced losses to insect predation, by 2009 87% of Indian cotton was GM. Though disputed the economic and environmental benefits of GM cotton in India to the individual farmer have been documented. In 2009, countries that grew 95% of the global transgenic crops were the United States (46%), Brazil (16%), Argentina (15%), India (6%), Canada (6%), China (3%), Paraguay (2%) and South Africa (2%). The Grocery Manufacturers of America estimate that 75% of all processed foods in the U.S. contain a GM ingredient.

In particular, Bt corn, which produces the pesticide within the plant itself, is widely grown, as are soybeans genetically designed to tolerate glyphosate herbicides. These constitute "input-traits" are aimed to financially benefit the producers, have indirect environmental benefits and marginal cost benefits to consumers. In the US, by 2009/2010, 93% of the planted area of soybeans, 93% of cotton, 86% of corn and 95% of the sugar beet were genetically modified varieties. Genetically modified soybeans carried herbicide-tolerant traits only, but maize and cotton carried both herbicide tolerance and insect protection traits (the latter largely the *Bacillus thuringiensis* Bt insecticidal protein). In the period 2002 to 2006, there were significant increases in the area planted to Bt protected cotton and maize, and herbicide tolerant maize also increased in sown area.

Legal Issues in the US

Alfalfa: On 21 June 2010, the US Supreme Court issued its first ruling in regard to a GM crop. This was a ruling in regard to Roundup Ready alfalfa. The case goes back to 2006, when organic farmers, concerned about the impact of GM alfalfa on their crops, sued Monsanto. In response, the California Northern District Court ruled that the United States Department of Agriculture (USDA) was in error when it approved the planting of Roundup Ready alfalfa.

According to the presiding judge, the law required the USDA to first conduct a full environmental study, which it had not done. It was the concern of the organic growers that the GM alfalfa could cross-pollinate with their organic alfalfa, making their crops unsalable in countries that forbid the growing of GM crops.

The impact of the current US Supreme Court ruling is somewhat unclear, with both sides appearing to claim victory. While Monsanto can claim technical victory in the case, various other issues still remain open, and will likely be litigated in the future. Meanwhile, the planting of GM alfalfa currently remains halted in the US, and it is unclear when it may resume.

Sugar Beets

Between 2009 and 2010, the United States District Court for the Northern District of California considered the case involving the planting of genetically modified sugar beets. This case involves Monsanto's breed of pesticide-resistant sugar beets.

Earlier in 2010, Judge Jeffrey S. White allowed the planting of GM sugar beets to continue, but he also warned that this may be blocked in the future while an environmental review was taking place. Finally, on 13 August 2010, Judge White ordered a halt to the planting of the genetically modified sugar beets in the US. He indicated that "the Agriculture Department had not adequately assessed the environmental consequences before approving them for commercial cultivation." The decision was the result of a lawsuit organised by the Centre for Food Safety, a US non-governmental organisation that is a critic of biotech crops.

Crop Yields

A 1999 study by Charles Benbrook, Chief Scientist of the Organic Centre, found that genetically engineered Roundup Ready soybeans did not increase yields. The report reviewed over 8,200 university trials in 1998 and found that Roundup Ready soybeans had a yield drag of 5.3% across all varieties tested. In addition, the same study found that farmers used 2-5 times more herbicide (Roundup) on Roundup Ready soybeans compared to other popular weed management systems.

However research published in Science in 2003 has shown that the use of genetically modified Bt cotton in India increased yields by 60% over the period 1998-2001 while the number of applications of insecticides against bollworm were three times less on average. A 2008 Soil Association report found that some scientific studies claimed that genetically modified varieties of plants do not produce higher crop yields than normal plants.

In 2009 the Union of Concerned Scientists summarized numerous peer-reviewed studies on the yield contribution of genetic engineering in the United States. This report examined the two most widely grown engineered crops—soybeans and maize (corn). Unlike many other studies, this work separated the yield contribution of the engineered gene from that of the many naturally occurring yield genes in crops.

The report found that engineered herbicide tolerant soy and maize did not increase yield at the national, aggregate level. Maize engineered with Bt insect resistance genes increased national yield by about 3 to 4 percent. Engineered crops increased net yield in all cases. The study concluded that in the United States, other agricultural methods have made a much greater contribution to national crop yield increases in recent years than genetic engineering. United States Department of Agriculture data record maize yield increases of about 28 percent since engineered varieties were first commercialized in the mid 1990s. The yield contribution of engineered genes has therefore been a modest fraction—about 14 percent—of the maize yield increase since the mid 1990s.

A 2010 article summarised the results of 49 peer reviewed studies on GM crops worldwide. On average, farmers in developed countries experienced increase in yield of 6% and in underdeveloped countries of 29%. Tillage was decreased by 25-58% on herbicide resistant soybeans, insecticide applications on Bt crops were reduced by 14-76% and 72% of farmers worldwide experienced positive economic results.

Coexistence and Traceability

The United States and Canada do not require labelling of genetically modified foods. However in certain other regions, such as the European Union, Japan, Malaysia and Australia, governments have required labelling so consumers can exercise choice between foods that have genetically modified, conventional or organic origins. This requires a labelling system as well as the reliable separation of GM and non-GM organisms at production level and throughout the whole processing chain.

For traceability, the OECD has introduced a "unique identifier" which is given to any GMO when it is approved. This unique identifier must be forwarded at every stage of processing. Many countries have established labelling regulations and guidelines on coexistence and traceability. Research projects such as Co-Extra, SIGMEA and Transcontainer are aimed at investigating improved methods for ensuring coexistence and providing stakeholders the tools required for the implementation of coexistence and traceability.

Detection

Testing on GMOs in food and feed is routinely done using molecular techniques like DNA microarrays or qPCR. These tests can be based on screening genetic elements (like p35S, tNos, pat, or bar) or event-specific markers for the official GMOs (like Mon810, Bt11, or GT73). The array-based method combines multiplex PCR and array technology to screen samples for different potential GMOs, combining different approaches (screening elements, plant-specific markers, and event-specific markers). The qPCR is used to detect specific GMO events by usage of specific primers for screening elements or event-specific markers. Controls are necessary to avoid false positive or false negative results. For example, a test for CaMV is used to avoid a false positive in the event of a virus contaminated sample.

PLU Codes

A Price Look-Up code beginning with the digit *8* indicates genetically modified food.

Genetically Modified Food Controversies

The genetically modified foods controversy is a dispute over the relative advantages and disadvantages of genetically modified (GM) food crops and other uses of genetically-modified organisms in food production. The dispute involves biotechnology companies, governmental regulators, non-governmental organizations and scientists. The dispute is most intense in Japan and Europe, where public concern about GM food is higher than in other parts of the world such as the United States. In the United States GM crops are more widely grown and the introduction of these products has been less controversial.

The five key areas of political controversy related to genetically engineered food are food safety, the effect on natural ecosystems, gene flow into non GE crops, moral/religious concerns, and corporate control of the food supply. To date, not a single instance of harm to human health or the environment has been documented with GM crops. Several benefits have been widely accepted and are uncontested in the scientific literature. These include reductions in insecticide use on GE cotton, enhanced biological diversity in GE cotton fields (compared to non-GE fields, enhanced farmer income and communal benefits, increased yields for poor farmers and improved health of farmworkers. Although the use of herbicide tolerant crops remain controversial, because of the need to spray herbicides, it is clear that the use of these crops has promoted a shift to less toxic herbicides.

The concept of substantial equivalence has been established to address the fears of some consumers about the safety of GE crops. "Substantial equivalence embodies the concept that if a new food or food component is found to be substantially equivalent to an existing food or food component, it can be treated in the same manner with respect to safety (i.e., the food or food component can be concluded to be as safe as the conventional food or food component)". The rationale for this approach is that it would be impossible to test all the new crop varieties every year for food safety. Only a few food crops on the market have been shown to cause adverse health effects and all of these were the result of conventional genetic modification, not from genetic engineering To date no adverse health effects caused by products approved for sale have been documented, although two products failed initial safety testing and were discontinued, due to allergic reactions. Most feeding trials have observed no toxic effects and saw that GM foods were equivalent in nutrition to unmodified foods, although a few non-peer-reviewed reports speculate physiological changes to GM food. Although there is now broad scientific consensus that GE crops on the market are safe to eat, some scientists and advocacy groups such as Greenpeace and World Wildlife Fund call for additional and more rigorous testing before marketing genetically engineered food.

Another area of controversy is what effect pest and herbicide-resistant crops have on ecosystems. For example, whereas 14 years of studies indicate cotton engineered for resistance to insects has resulted in massive reduction in insecticides globally, the use of crops engineered for herbicide tolerance has changed herbicide patterns, with a clear shift to use of less toxic herbicides. Studies assessing diversity of non-target insects has revealed enhanced biological diversity in GE fields compared to non-GE fields. The risk and effects of have also been cited as concerns, with the possibility that genes might spread from modified crops to wild relatives. There is no evidence that the risk of pollen transfer is greater for any of the GE crops on the market as compared to non-GE crops.

Health Risks and Benefits

Present Knowledge on GM Food Safety

Worldwide, there is a range of perspectives within non-governmental organizations on the safety of GM foods. For example, the US pro-GM group AgBioWorld has argued that GM foods have been proven safe, while other pressure groups and consumer rights groups, such as the Organic Consumers Association, and Greenpeace claim the long-term health risks which GM could pose, or the environmental risks associated with GM, have not yet been adequately investigated. In Japan, Consumers Union of Japan are opposed to GMO foods. They also claim that truly independent research in these areas is systematically blocked by the GM corporations which own the GM seeds and reference materials.

A 2008 review published by the Royal Society of Medicine noted that GM foods have been eaten by millions of people worldwide for over 15 years, with no reports of ill effects. Similarly a 2004 report from the US National Academies of Sciences stated: "To date, no adverse health effects attributed to genetic engineering have been documented in the human population." A 2004 review of feeding trials in the *Italian Journal of Animal Science* found no differences among animals eating genetically modified plants. A 2005 review in *Archives of Animal Nutrition* concluded that first-generation genetically modified foods had been found to be similar in nutrition and safety to non-GM foods, but noted that second-generation foods with "significant changes in constituents" would be more difficult to test, and would require further animal studies. However, a 2009 review in *Nutrition Reviews* found that although most studies concluded that GM foods do not differ in nutrition or cause any detectable toxic effects in animals, some studies did report adverse changes at a cellular level caused by some GM foods, concluding that "More scientific effort and investigation is needed to ensure that consumption of GM foods is not likely to provoke any form of health problem". A review published in 2009 by Dona and Arvanitoyannis concluded that "results of most studies with GM foods indicate that they may cause some common toxic effects such as hepatic, pancreatic, renal, or reproductive effects and may alter the hematological, biochemical, and immunologic parameters".

However responses to this review in 2009 and 2010 note that the Dona and Arvanitoyannis concentrated on articles with an anti-GM bias that have been refuted by scientists in peer-reviewed articles elsewhere- for example the 35S promoter, stability of transgenes, antibiotic marker genes and the claims for toxic effects of GM foods. In 2007, a review by Domingo of the toxicity by searching in the Publimed data base using 12 search terms, cited 68 references, found that the "number of references" on the safety of GM/transgenic crops was "surprisingly limited" and questioned whether the safety of genetically modified food has been demonstrated; the review also remarked that its conclusions were in agreement with three earlier reviews by Zdunczyk (2001), Bakshi (2003), and Pryme and Lembcke (2003).

However, an article in 2007 by Vain found 692 research studies focusing on GM crop and food safety and identified a strong increase in the publication of such articles in recent years. Vain commented that the multidisciplinarian nature of GM research complicates the retrieval of GM studies and requires using many search terms (he used more than 300) and multiple data bases.

Safety Assessments

The starting point for the safety assessment of genetically engineered food products is to assess if the food is "substantially equivalent" to its natural counterpart. To decide if a modified product is substantially equivalent, the product is tested by the manufacturer for unexpected changes in a limited set of components such as toxins, nutrients or allergens that are present in the unmodified food. If these tests show no significant difference between the modified and unmodified products, then no further food safety testing is required. The manufacturers data is then assessed by an independent regulatory body, such as the U.S. Food and Drug Administration.

However, if the product has no natural equivalent, or shows significant differences from the unmodified food, then further safety testing is carried out. A 2003 review in *Trends in Biotechnology* identified 7 main parts of a standard safety test:

1. Study of the introduced DNA and the new proteins or metabolites that it produces;
2. Analysis of the chemical composition of the relevant plant parts, measuring nutrients, anti-nutrients as well as any natural toxins or known allergens;
3. Assess the risk of gene transfer from the food to microorganisms in the human gut;
4. Study the possibility that any new components in the food might be allergens;
5. Estimate how much of a normal diet the food will make up;
6. Estimate any toxicological or nutritional problems revealed by this data;
7. Additional animal toxicity tests if there is the possibility that the food might pose a risk.

This process was examined further in a review published by Kuiper *et al.* 2002 in the journal *Toxicology*, which stated that substantial equivalence does not itself measure risks, but instead identifies differences between existing products and new foods, which might pose dangers to health. If differences do exist, identifying these differences is a starting point for a full safety assessment, rather than an end point.

The authors concluded that "The concept of substantial equivalence is an adequate tool in order to identify safety issues related to genetically modified products that have a traditional counterpart". However, the review also noted difficulties in applying this standard in practice, including the fact that traditional foods contain many chemicals that have toxic or carcinogenic effects and that our existing diets therefore have not been proven to be safe. This lack of knowledge on unmodified food poses a problem, as GM foods may have differences in anti-nutrients and natural toxins that have never been identified in the original plant, raising the possibility that harmful changes could be missed.

The application of substantial equivalence has also been more strongly criticized. For example, in a speech in 1999, Andrew Chesson of the University of Aberdeen, stated that substantial equivalence testing "could be flawed in some cases" and that some current safety tests could allow harmful substances to enter the human food chain. In a commentary in *Nature* Millstone *et al.* argued that all GM foods should have extensive biological, toxicological and immunological tests and that the concept of substantial equivalence based solely on chemical analyses of the components of a food should be abandoned.

They stated that this is necessary since it is currently impossible to predict the biological properties of a substance only from knowledge of its chemistry. This commentary was controversial and was criticized for misleading presentation of data and presenting an over-simplified version of safety assessments. For example, Kuiper *et al.* responded to this criticism by noting that equivalency testing does involve more than chemical tests and may include toxicity testing.

Medical writer Barbara Keeler and Marc Lappe argued in a 2001 article in the Los Angeles Times that the differences between Genetically Modified and conventional foods challenges the presumption of equivalence. Using Roundup ready soy that has been on the market since 1995 as an example, they noted the differences when compared to its unmodified counterpart.

Significantly lower levels of protein than unmodified soy. Significantly lower levels of phenylalanine, an essential amino acid and as a dietary supplement, the reason doctors advise the consumption of soy products. Levels of trypsin inhibitor were 27% higher and after toasting, Lectin was double that found in conventional soy, both are known allergens. GM soy also has 29% less choline, a B-complex vitamin. Round up ready soy had also stunted the growth of rats in Monsanto's study but had not effected cattle although it had increased the fat content of their milk. The authors do not maintain that modified soy is a hazard but that the FDA accepting such significant differences as being *substantially equivalent* illustrates the need for more rigorous testing, and preferably not by the biotech industries themselves. The value of current independent studies is problematic as, due to restrictive End-user agreements, researchers are forbidden by law from publishing independent research in peer reviewed journals without the approval of the agritech companies. Cornell University's Elson Shields, the spokesperson for a group of scientists who oppose this practice, submitted a statement to the United States Environmental Protection Agency (EPA) protesting that "as a result of restrictive access, no truly independent research can be legally conducted on many critical questions regarding the technology".

Scientific American noted that several studies that were initially approved by seed companies were later blocked from publication when they returned "unflattering" results. While recognising that seed companies' intellectual property rights need to be protected, Scientific American calls the practice dangerous and has called for the restrictions on research in the End-user agreements to be lifted immediately and for the EPA to require, as a condition of approval, that independent researchers have unfettered access to GM products for testing. The Welsh pressure group GM Free Cymru argues that governments should use independent studies rather than industry studies to assess crop safety. GM Free Cymru has also stated that independently funded researcher, Professor Bela Darvas of Debrecen University was refused Mon 863 Bt corn to use in his studies after previously publishing that a different variety of Monsanto corn was lethal to two Hungarian protected insect species and an insect classified as a rare.

Allergenicity

Worldwide, reports of allergies to all kinds of foods, particularly nuts, fish and shellfish, seem to be increasing, but it is not known if this reflects a genuine change in the risk of allergy, or an increased awareness of food allergies by the public. Some environmental organizations, such as the European Green Party and Greenpeace, have suggested that GM food might trigger food allergies, although other environmentalists have implicated causes as diverse as the greenhouse effect increasing pollen levels, greater exposure to synthetic chemicals, cleaner lifestyles, or more mold in buildings. A 2005 review in the journal *Allergy* of the results from allergen testing of current GM foods stated that "no biotech proteins in foods have been documented to cause allergic reactions".

A well-known case of a GM plant that did not reach the market due to it producing an allergic reaction was a new form of soybean intended for animal feed. The allergen was transferred unintentionally from the Brazil nut into genetically engineered soybeans, in a bid to improve soybean nutritional quality for animal feed use. This new protein increased the levels in the GM soybean of the natural essential amino acid methionine, which is commonly added to poultry feed. Investigation of the GM soybeans revealed that they produced immune reactions in people with Brazil nut allergies, since the methionine rich protein chosen by Pioneer Hi-Bred happened to be a major source of Brazil nut allergy. Although this soybean strain was not developed as a human food, Pioneer Hi-Bred discontinued further development of the GM soybean, due to the difficulty in ensuring that none of these soybeans entered the human food chain.

In November 2005 a pest-resistant field pea developed by the Australian CSIRO for use as a pasture crop was shown to cause an allergic reaction in mice. Work on this variety was immediately halted. The protein added to the pea did not cause the reaction in humans or mice in isolation, but when it was expressed in the pea, it exhibited a subtly different structure which may have caused the allergic reaction. The immunologist who tested the pea noted that crops need to be evaluated case-by-case, and a Greenpeace activist commented that to his knowledge, no countries required feeding studies for approval of genetically modified foods.

Plant scientist Maarten J Chrispeels has made these comments about this example:

The recent Prescott et al. paper in JFAC contains a very interesting study on the immunogenicity of amylase [starch digestion enzyme] inhibitor in its native form (isolated from beans) and expressed as a transgene in peas. First of all, amylase inhibitor is a food protein, but also a "toxic" protein because it inhibits our digestive amylases. This is one of the reasons you have to cook your beans! (The other toxic bean protein is phytohemagglutinin and it is much more toxic).

This particular amylase inhibitor is found in the common bean (other species have other amylase inhibitors). Even though it is a food protein, it is unlikely ever to be used for genetic engineering of human foods because it inhibits our amylases. What the results show is that the protein, when synthesized in pea cotyledons has a different immunogenicity than when it is isolated from bean cotyledons (the native form). This is somewhat surprising but may be related to the presence of slightly different carbohydrate chains.

These cases of products that failed safety testing can either be viewed as evidence that genetic modification can produce unexpected and dangerous changes in foods, or alternatively that the current tests are effective at identifying any safety problems before foods come on the market.

Genetic modification can also be used to remove allergens from foods, which may, for example, allow the production of soy products that would pose a smaller risk of food allergies than standard soybeans. A hypo-allergenic strain of soybean was tested in 2003 and shown to lack the major allergen that is found in the beans. A similar approach has been tried in ryegrass, which produces pollen that is a major cause of hayfever: here a fertile GM grass was produced that lacked the main pollen allergen, demonstrating that the production of hypoallergenic grass is also possible.

Environmental Risks and Benefits

The large scale growth of GM plants may have both positive and negative effects on the environment. These may be both direct effects, on organisms that feed on or interact with the crops, or wider effects on food chains produced by increases or decreases in the numbers of other organisms. As an example of benefits, insect-resistant Bt-expressing crops will reduce the number of pest insects feeding on these plants, but as there are fewer pests, farmers do not have to apply as much insecticide, which in turn tends to increase the number of non-pest insects in these fields. A 2006 study of the global impact of GM crops, published by the UK consultancy PG Economics and funded by Illinois-Missouri Biotechnology Alliance, concluded that globally, the technology reduced pesticide spraying by 286,000 tons in 2006, decreasing the environmental impact of herbicides and pesticides by 15%.

By reducing the amount of ploughing needed, GM technology led to reductions of greenhouse gases from soil equivalent to removing 6.56 million cars from the roads. However, a 2009 study published by the Organic Centre stated that the use of genetically engineered corn, soybean, and cotton increased the use of herbicides by 383 million pounds (191,500 tons), and pesticide use by 318.4 million pounds (159,200 tons). As an example of a concern about environmental risk, a lab at Cornell University published an article which caused worry in the US that Bt-corn pollen might affect the monarch butterfly. However this concern was disproved by six comprehensive articles in the Proceedings of the National Academy of Sciences in 2001. Monarch populations increased, despite increased Bt corn probably due to reduced pesticide use. Other possible effects might come from the spread of genes from modified plants to unmodified relatives, which might produce species of weeds resistant to herbicides. In some areas of the US "superweeds" have evolved naturally, these weeds are resistant to herbicides and have forced farmers to return to traditional crop management practices.

There has been controversy over the results of a farm-scale trial in the United Kingdom comparing the impact of GM crops and conventional crops on farmland biodiversity. Some claimed that the results showed that GM crops had a significant negative impact on wildlife. They pointed out that the studies showed that using herbicide resistant GM crops allowed better weed control and that under such conditions there were fewer weeds and fewer weed seeds. This result was then extrapolated to suggest that GM crops would have significant impact on the wildlife that might rely on farm weeds. The President of the Royal Society, the body that had carried out the trials, stated that "To generalize and declare 'all GM is bad' or 'all GM is good' for the environment as a result of these experiments is a gross over-simplification", arguing that although the trials showed that the combination of some GM crops with long-lasting herbicides were bad for biodiversity, using other GM crops without these herbicides increased biodiversity.

In July 2005 British scientists showed that transfer of a herbicide-resistance gene from GM oilseed rape to a wild cousin, charlock, and wild turnips was possible.

Many agricultural scientists and food policy specialists view GM crops as an important element in sustainable food security and environmental management. This point of view is summarized in the ABIC Manifesto:

On our planet, 18% of the land mass is used for agricultural production. This fraction cannot be increased substantially. It is absolutely essential that the yield per unit of land increases beyond current levels given that: The human population is still growing, and will reach about nine billion by 2040;70,000 km^2 of agricultural land (equivalent to 60% of the German agricultural area) are lost annually to growth of cities and other non-agricultural uses; Consumer diets in developing countries are increasingly changing from plant-based proteins to animal protein, a trend that requires a greater amount of crop-based feeds.

Other scientists, such as Dr. Charles Benbrook, argue that improvement of global food security is hardly being addressed by genetic research and that a lack of yield is often not caused by insufficient genetic resources. Regarding the issues of intellectual property and patent law, an international report from the year 2000 states:

If the rights to these tools are strongly and universally enforced-and not extensively licensed or provided pro bono in the developing world- then the potential applications of GM technologies described previously are unlikely to benefit the less developed nations of the world for a long time (ie until after the restrictions conveyed by these rights have expired).

Issues with Bt Maize

A well publicized claim associated with Bt crops (or transgenic maize) was the concern that pollen from Bt maize might kill the monarch butterfly. This report was puzzling because the pollen from most maize hybrids contains much lower levels of Bt than the rest of the plant and led to multiple follow-up studies. One possible issue revealed in these studies is the possibility that the initial study was flawed; based on the way the pollen was collected, in that they collected and fed non-toxic pollen that was mixed with anther walls that did contain Bt toxin. A collaborative research exercise was carried out over two years by several groups of scientists in the US and Canada, looking at the effects of Bt pollen in both the field and the laboratory. This resulted in a risk assessment that concluded that any risk posed by the corn to butterfly populations under real-world conditions was negligible. The USDA has stated that the weight of the evidence is that Bt crops do not pose a risk to the monarch butterfly. An independent 2002 review of the scientific literature concluded that "the commercial large-scale cultivation of current Bt–maize hybrids did not pose a significant risk to the monarch population" and noted that despite large-scale planting of GM crops, the butterfly's population is increasing.

In 2007 Andreas Lang, Éva Lauber and Bela Darvas criticized these studies, arguing that there can be a great difference in the effects between the acute exposure tested for and chronic exposure. Moreover, they stated that the "worst case conditions" performed were not in fact worst case scenarios, as laboratory conditions with ample food supply and a favourable climate ensure healthy subjects. They instead believe that in the wild, low temperatures, rain and parasites and disease might exacerbate a Bt effect on butterfly larvae. Their own experiments suggested that some butterfly species were negatively affected by such chronic exposure. Jorg Romeis, who conducted the original studies, replied that if species of Butterfly are affected as Darvas claims that a "more comprehensive assessment will be needed and, depending on the degree and nature of concern, this may extend to field testing".

A 2001 report in *Nature* presented evidence that Bt maize was cross-breeding with unmodified maize in Mexico, although the data in this paper was later described as originating from an artifact and *Nature* stated that "the evidence available is not sufficient to justify the publication of the original paper". A subsequent large-scale study, in 2005, failed to find any evidence of contamination in Oaxaca. However, other authors have stated that they also found evidence of cross-breeding between natural maize and transgenic maize.

There is also a risk that for example, transgenic maize will crossbreed with wild grass variants, and that the Bt-gene will end up in a natural environment, retaining its toxicity. An event like this would have ecological implications. However, there is no evidence of crossbreeding between maize and wild grasses.

In 2009 it was reported that 82,000 hectares (200,000 acres) of Bt corn in South Africa failed to produce seeds. Monsanto claimed average yield was reduced by 25% in those fields affected, it compensated the farmers concerned and the corn varieties were affected by a mistake made in the seed breeding process. Marian Mayet an environmental activist and director of the Africa-centre for biosecurity in Johannesburg called for a government investigation and asserted that the biotechnology was at fault, *"You cannot make a mistake with three different varieties of corn"*. In 2009 South African farmers planted 1,900,000 hectares (4,700,000 acres) of GM maize (73% of the total crop).

As of 2007, a phenomenon called Colony Collapse Disorder (CCD) was noticed in bee hives all over North America, and elsewhere. Although it is not certain if this is a new phenomenon, initial ideas on the possible causes ranged from poor nutrition, infections, parasites and pesticide use. More unusual speculations included radio waves from cellphone base stations, climate change, and the use of transgenic crops containing Bt. The Mid-Atlantic Apiculture Research and Extension Consortium published a report on 2007-03-27 that found no evidence that pollen from Bt crops is adversely affecting bees. Several researchers in the US have since attributed CCD to the spread of a new virus called Israeli acute paralysis virus, although other parasites have also been implicated.

Legal Issues in the US

Alfalfa

On 21 June 2010, the US Supreme Court in *Monsanto v. Geertson Seed Farms* issued its first ruling about GM crops. This case relates to Roundup Ready alfalfa and dates back to 2006, when a coalition of groups led by the Centre for Food Safety raised concerns about environmental impacts that the United States Department of Agriculture (USDA) allegedly failed to address before approving the planting of Roundup Ready alfalfa. In response, in 2007, Judge Charles Breyer of the California Northern District Court ruled that USDA was in error when it approved the planting of Roundup Ready alfalfa.

In June 2009, a divided three-judge panel on the 9th U.S. Circuit Court of Appeals upheld Judge Charles Breyer's decision, and Monsanto appealed to the Supreme Court.

The impact of the current US Supreme Court ruling is somewhat unclear, with both sides appearing to claim victory.

Sugar Beets

There is also a somewhat similar case involving sugar beets before the same California Northern District Court. This is a case involving Monsanto's breed of pesticide-resistant sugar beets. While Judge Jeffrey S. White (issuing his ruling in the spring of 2010) allowed the planting of GM sugar beets to continue, he also warned that this may be blocked in the future while an environmental review is taking place.

Finally, on 13 August 2010, Judge Jeffrey S. White ordered the halt to the planting of the genetically modified sugar beets in the US.

He indicated that "the Agriculture Department had not adequately assessed the environmental consequences before approving them for commercial cultivation."

In 2010, before the ruling, 95% of the sugar beet grown in the US was GM. About half the sugar supply in the US came from sugar beet.

Control of the Market

Patent holder companies use their control of their own GMO to corner the market and gain profit. However, other companies often compete for the little market share available to GM foods worldwide. Detractors such as Greenpeace say that patent rights give corporations a dangerous amount of control over their product while corporations claim that they need product control in order to prevent seed piracy, fulfil financial obligations to shareholders and invest in further GM development. Governments have also sought to protect their commercial interests through punitive measures against countries resisting GM foods on ethical or scientific grounds: after moves in France to ban a Monsanto GM corn variety, the US embassy recommended that 'we calibrate a target retaliation list that causes some pain across the EU'.

Controversial Cases

Pusztai Potato Controversy

A large media event occurred in 1998 when scientist Arpad Pusztai, who "was considered a world expert on plant lectins", reported in a television interview that he had found that rats fed potatoes genetically modified by the English biotech company *Cambridge Agricultural Genetics* to contain lectin, a natural insecticide in snowdrop plants, had stunted growth and immune system damage. Initially the Rowett Institute's director, Philip James, congratulated Pusztai after the interview, but his attitude changed after confusion arose over whether the lectin was from a jackbean or a snowdrop (the jackbean lectin is poisonous to the intestines). A press release from the Rowett Institute stated that the lectin had come from a jackbean, but no experiments were made using lectins from the jackbean; in the confusion, it was alleged that Pusztai hadn't actually carried out the experiments. James suspended Pusztai and used misconduct procedures to seize the data. His annual contract was not renewed and both Pusztai and his wife were banned from speaking publicly. In October 1998 the Rowett Institute published an audit criticizing Pusztai's results, which, along with Pusztai's raw data, was sent to six anonymous reviewers who criticized Pusztai's work. Pusztai responded that the raw data was "never intended for publication under intense scrutiny". Pusztai sent the audit report and his rebuttal to scientists who requested it, and in February 1999, twenty-one European and American scientists, mostly friends or acquaintances of Pusztai, orchestrated by the environmental group Friends of the Earth, released a memo supporting Pusztai. Stanley Ewen, who worked with Pusztai, conducted a followup study supporting Pusztai's work and presented the work to a lectin meeting in Sweden.

In October 1999 Pusztai's research was published in the journal The Lancet. Because of the controversial nature of his research the data in this paper, co-authored by Stanley Ewen, was seen by a total of six reviewers when presented for peer review; five of these reviewers judged the work acceptable, although one of the five "deemed the study flawed but favoured publication to avoid suspicions of a conspiracy against Pusztai and to give colleagues a chance to see the data for themselves". The paper did not mention stunted growth or immunity issues, but reported that rats fed on potatoes genetically modified with the snowdrop lectin had "thickening in the mucosal lining of their colon and their jejunum" when compared with rats fed on non modified potatoes. Three Dutch scientists criticized the study on the grounds that the unmodified potatoes were not a fair control diet, and that any rats fed only on potatoes will suffer from a protein deficiency; Pusztai responded to these criticisms by stating that the protein and energy were comparable, and that "a sample size of six is perfectly normal in studies like this".

The genetically modified potatoes were later destroyed by *Cambridge Agricultural Genetics* as were all details of their modification.

Bioequivalence Study of a Corn Cultivar

A controversy arose around biotech company Monsanto's data on a 90-Day Rat Feeding Study on the MON863 strain of GM corn. In May 2005, critics of GM foods pointed to differences in kidney size and blood composition found in this study, suggesting that the observed differences raises questions about the regulatory concept of substantial equivalence. Anti-GM campaigner Jeffrey M. Smith, writing in Biophile Magazine, quoting comments from Pusztai and Seralini, has stated that nutritional studies typically use young, fast-growing animals with starting weights not varying by more than 2% from the average whereas Monsanto's research design used a mix of young and old animals with starting weights ranging from 198.4 to 259.8 grams. Seralini and two other authors published a study of these data, funded by Greenpeace, in 2007 making similar points.

The raising of this issue prompted the European Food Safety Authority (EFSA) to reexamine the safety data on this strain of corn. The EFSA concluded that the observed small numerical decrease in rat kidney weights were not biologically meaningful, and the weights were well within the normal range of kidney weights for control animals. There were no corresponding microscopic findings in the relevant organ systems, and they stated that all blood chemistry and organ weight values fell within the "normal range of historical control values" for rats. In addition the EFSA review stated that the statistical methods used by Seralini et al. in the analysis of the data were incorrect. The European Commission has approved the ÌÍÍ863 corn for animal and human consumption. Food Standards Australia New Zealand reviewed the 2007 Seralini et al study and concluded that "...all of the statistical differences between rats fed MON 863 corn and control rats are attributable to normal biological variation."

Greenpeace stated in a 2007 press release that Seralini et al. had completed a similar analysis of the NK603 strain of corn and came to similar conclusions as they did in their previous study. Seralini et al included this in a re-analysis of three existing rat feeding studies published in 2009.

The European Food Safety Authority reviewed the 2009 Seralini et al paper and concluded that the author's claims were not supported by the data in their paper. They noted that many of their fundamental statistical chriticisms of the 2007 paper also applied to the 2009 paper. There was no new information that would change the EFSA's conclusions that the three GM maize types were safe for human, animal health and the environment The French High Council of Biotechnologies Scientific Committee (HCB) also reviewed the 2009 study and concluded that it "..presents no admissible scientific element likely to ascribe any haematological, hepatic or renal toxicity to the three re-analysed GMOs." The HCB also questioned the author's independence. Food Standards Australia New Zealand concluded that the results from the 2009 Seralini et al. study were due to chance alone.

Contamination Issues

In the 1990s genetically modified Flax tolerant to herbicide residues in soil was developed by the Crop Development Centre (CDC) at the

University of Saskatchewan in Canada. Named Flax variety FP967, but commonly called *CDC Triffid*, research was controversially halted following protests from Canadian farmers who stood to lose up to 70% of their traditional export markets if it was introduced. GM Flax was deregistered, its sale was criminalized and in 2001 all modified seeds were destroyed. No modified crops had been planted and no seed had been sold but GM industry proponent Alan McHughen controversially passed out sample packets of seeds at presentations. In early September 2009, Flax imported into Germany was found to be contaminated with *CDC Triffid* causing the price of Canadian Flax to fall 32 percent. By mid November 35 countries reported contamination of imported Canadian Flax which has now been banned by the European Union. Canadian farmers are expected to be responsible for the cost of the cleanup and testing of future crops.

In 2000, Aventis StarLink corn, which had been approved only as animal feed due to concerns about possible allergic reactions in humans, was found contaminating corn products in U.S. supermarkets. An episode involving Taco Bell taco shells was particularly well publicized which resulted in sales of StarLink seed being discontinued. The registration for the Starlink varieties was voluntarily withdrawn by Aventis in October 2000. Aid sent by the UN and the US to Central African nations was also found to be contaminated with StarLink corn and the aid was rejected. The US corn supply has been monitored for Starlink Bt proteins since 2001 and no positive samples have been found since 2004.

Public Perception

Research by the Pew Initiative on Food and Biotechnology has shown that in 2005 Americans' knowledge of genetically modified foods and animals continues to remain low, and their opinions reflect that they are particularly uncomfortable with animal cloning. In one instance of consumer confusion, DNA Plant Technology's Fish tomato transgenic organism was conflated with Calgene's Flavr Savr transgenic food product. The Pew survey also showed that despite continuing concerns about GM foods, American consumers do not support banning new uses of the technology, but rather seek an active role from regulators to ensure that new products are safe.

Only 2% of Britons were said to be "happy to eat GM foods", and more than half of Britons were against GM foods being available to the public, according to a 2003 study. However a 2009 review article of European consumer polls concluded that opposition to GMOs in

Europe has been gradually decreasing. Approximately half of European consumers accepted gene technology, particularly when benefits for consumers and for the environment could be linked to GMO products. 80 % of respondents did not cite the application of GMOs in agriculture as a significant environmental problem. Many consumers seem unafraid of health risks from GMO products and most European consumers did not actively avoid GMO products while shopping.

In Australia, GM foods that have novel DNA, novel protein, altered characteristics or has to be cooked or prepared in a different way compared to the conventional food have, since December 2001, had to be identified on food labels. However, multiple surveys have shown that while 45% of the public will accept GM foods, some 93% demand all genetically modified foods be labelled as such. A 2007 survey by the Food Standards Australia and New Zealand found that 27% of Australians looked at the label to see if it contained GM material when purchasing a grocery product for the first time.

Labelling legislation has been introduced and rejected several times since 1996 on the grounds of "restraint of trade" due to the cost of labelling. The controversy erupted again in 2009 when Graincorp, the nations largest grain handler, announced it would mix GM Canola with its unmodified grain. Traditional growers, who largely rely on GM-free markets, had been told they would need to pay to have their produce certified GM free. Graincorp reversed its decision the same year. Critics such as Greenpeace and the Gene Ethics Network have renewed calls for more labelling.

Opponents of genetically modified food often refer to it as "Frankenfood", after Mary Shelley's character Frankenstein and the monster he creates, in her novel of the same name. The term was coined in 1992 by Paul Lewis, an English professor at Boston College who used the word in a letter he wrote to the *New York Times* in response to the decision of the US Food and Drug Administration to allow companies to market genetically modified food. The term "Frankenfood" has become a battle cry of the European side in the US-EU agricultural trade war.

Critics have protested in regards to the appointment of pro GM lobbyists to senior positions in the FDA. Michael R. Taylor has been appointed as a senior adviser to the FDA on food safety and Dennis Wolff is expected to take up the position of Under-Secretary of the newly created Agriculture for Food Safety.

Taylor is a former Monsanto lobbyist credited as being responsible for the implementation of "substantial equivalence" in place of food safety studies and for his advocacy that resulted in the Delaney clause that prohibited the inclusion of "any chemical additive found to induce cancer in man.. or animals" in processed foods being amended in 1996 to allow the inclusion of pesticides in GMOs. Wolff is the Pennsylvania Secretary of Agriculture who successfully lobbied to ban organic farmers from labelling their products as being GM free and was a proponent of the "ACRE" initiative which gave the Pennsylvania state attorney general's office the authority to sue municipalities that banned GMOs. Several anti-GMO organisations have organised petitions demanding Taylor's resignation and opposing Wolff's appointment and also conducted letter writing campaigns protesting the conflict of interest.

Religious Issues

As of yet, no GM foods have been designated as unacceptable by religious authorities.

Economic and environmental effects:

- Many proponents of genetically engineered crops claim they lower pesticide usage and have brought higher yields and profitability to many farmers, including those in developing nations. For example, a 2010 study by US scientists, found that the economic benefit of Bt corn to farmers in five mid-west states was $6.9 billion over the previous 14 years. They were surprised that the majority ($4.3 billion) of the benefit accrued to non-Bt corn. This was speculated to be because the European Corn Borers that attack the Bt corn die and there are fewer left to attack the non-GM corn nearby.

- The United States has seen a widespread adoption of genetically-engineered corn, cotton and soybean crops since 1996.

- In 2010, the U.S. National Academy of Sciences reported that genetically engineered crops had resulted in reduced pesticide application and reduced soil erosion from tilling. The report also stated that the advent of glyphosate-herbicide resistant weeds—that have developed because of the use of engineered crops—could cause the genetically engineered crops to lose their effectiveness unless farmers also use other established weed management strategies.

- In a study by Scientists at the University of Arkansas published in 2010 showed that about 83 percent of wild or weedy canola they tested contained genetically modified herbicide resistance genes, and they also found some plants that contained resistance to both herbicides, a combination of transgenic traits that had not been developed in canola crops. That leads us to believe that these wild populations that contain modified genes have become established populations.

Bans

- In 2002, Zambia cut off the flow of Genetically Modified Food (mostly maize) from UN's World Food Programme. This left a famine-stricken population without food aid.

- In December 2005 the Zambian government changed its mind in the face of further famine and allowed the importation of GM maize. However, the Zambian Minister for Agriculture Mundia Sikatana has insisted that the ban on genetically modified maize remains, saying "We do not want GM (genetically modified) foods and our hope is that all of us can continue to produce non-GM foods."

- In April 2004 Hugo Chavez announced a total ban on genetically modified seeds in Venezuela.

- In January 2005, the Hungarian government announced a ban on importing and planting of genetic modified maize seeds, which was subsequently authorized by the EU.

- On August 18, 2006, American exports of rice to Europe were interrupted when much of the U.S. crop was confirmed to be contaminated with unapproved engineered genes, possibly caused by cross-pollination with conventional crops.

- On February 9, 2010, Indian Environment Minister, Jairam Ramesh, imposed a moratorium on the cultivation of GMF "for as long as it is needed to establish public trust and confidence". His decision was made after protest from several groups responding to regulatory approval of the cultivation of Bt brinjal, a GM eggplant in October, 2009.

U.S. Government Reaction to European Ban

In recent years, France and several other European countries banned Monsanto's MON-810 corn and similar genetically modified food crops. In late 2007, the U.S. ambassador to France recommended "moving to retaliation" against France and the European Union in an attempt to fight the French ban and changes in European policy toward genetically modified crops, according to a U.S. government diplomatic cable obtained by WikiLeaks. The U.S. ambassador to France recommended retaliation to cause "some pain across the EU."

Intellectual Property

Traditionally, farmers in all nations saved their own seed from year to year. It should be noted that this does not apply in more agriculturally developed countries for some crops. Corn is one example where producers generally have not saved seed since the early 1900s with the advent of hybrid corn through selective breeding. Seed producers grow the seed corn instead due to the effort needed to produce hybrids. The offspring of the hybrid corn, while still viable, lose the beneficial traits of the parents, resulting in the loss of hybrid vigor. In these cases, the use of hybrid plants has been the primary reason for growers not saving seed, not intellectual property issues, and has been in practice well before genetically-modified seed was developed. However, the practice of not saving seed in non-hybrid crops, such as soybean, is mainly due to intellectual property regulations. Allowing to follow this practice with genetically modified seed would result in seed developers losing the ability to profit from their breeding work. Therefore, genetically-modified seed is subject to licensing by their developers in contracts that are written to prevent farmers from following this practice.

Enforcement of patents on genetically modified plants is often contentious, especially because of gene flow. In 1998, 95-98 percent of about 10 km planted with canola by Canadian farmer Percy Schmeiser were found to contain Monsanto Company's patented Roundup Ready gene although Schmeiser had never purchased seed from Monsanto. The initial source of the plants was undetermined, and could have been through either gene flow or intentional theft. However, the overwhelming predominance of the trait implied that Schmeiser must have intentionally selected for it. The court determined that Schmeiser had saved seed from areas on and adjacent to his property where Roundup had been sprayed, such as ditches and near power poles.

Although unable to prove direct theft, Monsanto sued Schmeiser for piracy since he knowingly grew *Roundup Ready* plants without paying royalties (Ibid). The case made it to the Canadian Supreme Court, which in 2004 ruled 5 to 4 in Monsanto's favour. The dissenting judges focused primarily on the fact that Monsanto's patents covered only the gene itself and glyphosate resistant *cells*, and failed to cover transgenic plants in their entirety. All of the judges agreed that Schmeiser would not have to pay any damages since he had not benefited from his use of the genetically modified seed. In response to criticism, Monsanto Canada's Director of Public Affairs stated that "It is not, nor has it ever been Monsanto Canada's policy to enforce its patent on Roundup Ready crops when they are present on a farmer's field by accident...Only when there has been a knowing and deliberate violation of its patent rights will Monsanto act."

Future Developments

Future envisaged applications of GMOs are diverse and include drugs in food, bananas that produce human vaccines against infectious diseases such as Hepatitis B, metabolically engineered fish that mature more quickly, fruit and nut trees that yield years earlier, foods no longer containing properties associated with common intolerances, and plants that produce new plastics with unique properties. While their practicality or efficacy in commercial production has yet to be fully tested, the next decade may see exponential increases in GM product development as researchers gain increasing access to genomic resources that are applicable to organisms beyond the scope of individual projects. Safety testing of these products will also, at the same time, be necessary to ensure that the perceived benefits will indeed outweigh the perceived and hidden costs of development. Plant scientists, backed by results of modern comprehensive profiling of crop composition, point out that crops modified using GM techniques are less likely to have unintended changes than are conventionally bred crops.

Health Risks

In the United States, the FDA Centre for Food Safety and Applied Nutrition reviews summaries of food safety data developed and voluntarily submitted by developers of engineered foods, in part on the basis of comparability to conventionally-produced foods. There are no specific tests required by FDA to determine safety. FDA does not approve the safety of engineered foods, but after its review, acknowledges that the developer of the food has asserted that it is safe. A 2008 review published by the Royal Society of Medicine noted that GM foods have been eaten by millions of people worldwide for over 15 years, with no reports of ill effects. Similarly a 2004 report from the US National Academies of Sciences stated: "To date, no adverse health effects attributed to genetic engineering have been documented in the human population."

There have, however, been no epidemiological studies to determine whether engineered crops have caused any harm to the public. Without such studies, it is unlikely that harm, if it occurred, would be detected or attributed to engineered foods. Worldwide, there are a range of perspectives within non-governmental organizations on the safety of GM foods.

For example, the US pro-GM pressure group AgBioWorld has argued that GM foods have been proven safe, while other pressure groups and consumer rights groups, such as the Organic Consumers Association, and Greenpeace claim the long term health risks which GM could pose, or the environmental risks associated with GM, have not yet been adequately investigated.

In 1998 Rowett Research Institute scientist Arpad Pusztai reported that consumption of potatoes genetically modified to contain lectin had adverse intestinal effects on rats. Pusztai eventually published a paper, co-authored by Stanley Ewen, in the journal, The Lancet. The paper claimed to show that rats fed on potatoes genetically modified with the snowdrop lectin had unusual changes to their gut tissue when compared with rats fed on non modified potatoes.

The experiment modified potatoes to add a toxin (snowdrop lectin), but the experiment failed to include a control for the toxin alone or a control for genetic modifications alone (without added toxin); therefore, no conclusion could be made about the safety of the genetic engineering. The experiment has been criticised by other scientists on the grounds that the unmodified potatoes were not a fair control diet and that all the rats may have been sick, due to them being fed a diet of only potatoes.

In 2009 three scientists (Vendomois et al) published a statistical re-analysis of three feeding trials that had previously been published by others as establishing the safety of genetically modified corn. The new article claimed that their statistics instead showed that the three patented crops (Mon 810, Mon 863, and NK 603) developed and owned by Monsanto cause liver, kidney, and heart damage in mammals. A 2007 analysis of part of this data by the same group of scientists funded by Greenpeace was assessed by a panel of independent toxicologists in a study funded by Monsanto and published in the journal *Food and chemical toxicology*.

The reviewers reported that the study was statistically flawed and providing no evidence of adverse effects. The French High Council of Biotechnologies Scientific Committee reviewed the 2009 Vendomois et al study and concluded that it "..presents no admissible scientific element likely to ascribe any haematological, hepatic or renal toxicity to the three re-analysed GMOs." An evaluation by the European Food Safety Authority of the 2009 and 2007 studies noted that most of the results were within natural variation and they did not consider any of the effects reported biologically relevant. A review by Food Standards Australia New Zealand of the 2009 Vendomois et al study concluded that the results were due to chance alone.

Gene Transfer

As of January 2009 there has only been one human feeding study conducted on the effects of genetically modified foods. The study involved seven human volunteers who had previously had their large intestines removed. These volunteers were to eat GM soy to see if the DNA of the GM soy transferred to the bacteria that naturally lives in the human gut. Researchers identified that three of the seven volunteers had transgenes from GM soya transferred into the bacteria living in their gut before the start of the feeding experiment. As this low-frequency transfer did not increase after the consumption of GM Soy, the researchers concluded that gene transfer did not occur during the experiment. In volunteers with complete digestive tracts, the transgene did not survive passage through intact gastrointestinal tract. Anti-GM advocates believe the study should prompt additional testing to determine its significance.. Other studies have found DNA from M13 virus, GFP and even ribulose-1,5-bisphosphate carboxylase (Rubisco) genes in the blood and tissue of ingesting animals.

Two studies on the possible effects of feeding genetically modified feeds to animals found that there was no significant differences in the safety and nutritional value of feedstuffs containing material derived from genetically modified plants. Specifically, the studies noted that no residues of recombinant DNA or novel proteins have been found in any organ or tissue samples obtained from animals fed with GMP plants.

Allergies

In the mid 1990s Pioneer Hi-Bred tested the allergenicity of a transgenic soybean that expressed a Brazil nut seed storage protein in hope that the seeds would have increased levels of the amino acid methionine. The tests (radioallergosorbent testing, immunoblotting, and skin-prick testing) showed that individuals allergic to Brazil nuts were also allergic to the new GM soybean. Pioneer has indicated that it will not develop commercial cultivars containing Brazil nut protein because the protein is likely to be an allergen.

Traceability

In a January 2010 paper by Costa et al. the extraction and detection of DNA along a complete industrial soybean oil processing chain was described to monitor the presence of Roundup Ready (RR) soybean: "The amplification of soybean lectin gene by end-point polymerase chain reaction (PCR) was successfully achieved in all the steps of extraction and refining processes, until the fully refined soybean oil. The amplification of RR soybean by PCR assays using event-specific primers was also achieved for all the extraction and refining steps, except for the intermediate steps of refining (neutralisation, washing and bleaching) possibly due to sample instability. The real-time PCR assays using specific probes confirmed all the results and proved that it is possible to detect and quantify genetically modified organisms in the fully refined soybean oil. To our knowledge, this has never been reported before and represents an important accomplishment regarding the traceability of genetically modified organisms in refined oils."

Reduced Vulnerability of Crops to Environmental Stresses

Crops containing genes that will enable them to withstand biotic and abiotic stresses may be developed. For example, drought and excessively salty soil are two important limiting factors in crop productivity. Biotechnologists are studying plants that can cope with these extreme conditions in the hope of finding the genes that enable them to do so and eventually transferring these genes to the more desirable crops.

One of the latest developments is the identification of a plant gene, At-DBF2, from Arabidopsis thaliana, a tiny weed that is often used for plant research because it is very easy to grow and its genetic code is well mapped out. When this gene was inserted into tomato and tobacco cells, the cells were able to withstand environmental stresses like salt, drought, cold and heat, far more than ordinary cells. If these preliminary results prove successful in larger trials, then At-DBF2 genes can help in engineering crops that can better withstand harsh environments. Researchers have also created transgenic rice plants that are resistant to rice yellow mottle virus (RYMV). In Africa, this virus destroys majority of the rice crops and makes the surviving plants more susceptible to fungal infections.

Increased Nutritional Qualities

Proteins in foods may be modified to increase their nutritional qualities. Proteins in legumes and cereals may be transformed to provide the amino acids needed by human beings for a balanced diet. A good example is the work of Professors Ingo Potrykus and Peter Beyer in creating Golden rice.

Improved Taste, Texture or Appearance of Food

Modern biotechnology can be used to slow down the process of spoilage so that fruit can ripen longer on the plant and then be transported to the consumer with a still reasonable shelf life. This alters the taste, texture and appearance of the fruit. More importantly, it could expand the market for farmers in developing countries due to the reduction in spoilage. However, there is sometimes a lack of understanding by researchers in developed countries about the actual needs of prospective beneficiaries in developing countries. For example, engineering soybeans to resist spoilage makes them less suitable for producing tempeh which is a significant source of protein that depends on fermentation. The use of modified soybeans results in a lumpy texture that is less palatable and less convenient when cooking.

The first genetically modified food product was a tomato which was transformed to delay its ripening. Researchers in Indonesia, Malaysia, Thailand, Philippines and Vietnam are currently working on delayed-ripening papaya in collaboration with the University of Nottingham and Zeneca. Biotechnology in cheese production: enzymes produced by micro-organisms provide an alternative to animal rennet – a cheese coagulant – and an alternative supply for cheese makers. This also eliminates possible public concerns with animal-derived material, although there are currently no plans to develop synthetic milk, thus making this argument less compelling. Enzymes offer an animal-friendly alternative to animal rennet. While providing comparable quality, they are theoretically also less expensive.

About 85 million tons of wheat flour is used every year to bake bread. By adding an enzyme called maltogenic amylase to the flour, bread stays fresher longer. Assuming that 10–15% of bread is thrown away as stale, if it could be made to stay fresh another 5–7 days then perhaps 2 million tons of flour per year would be saved. Other enzymes can cause bread to expand to make a lighter loaf, or alter the loaf in a range of ways.

Reduced Dependence on Fertilizers, Pesticides and other Agrochemicals

Most of the current commercial applications of modern biotechnology in agriculture are on reducing the dependence of farmers on agrochemicals. For example, *Bacillus thuringiensis* (Bt) is a soil bacterium that produces a protein with insecticidal qualities. Traditionally, a fermentation process has been used to produce an insecticidal spray from these bacteria. In this form, the Bt toxin occurs as an inactive protoxin, which requires digestion by an insect to be effective. There are several Bt toxins and each one is specific to certain target insects.

Crop plants have now been engineered to contain and express the genes for Bt toxin, which they produce in its active form. When a susceptible insect ingests the transgenic crop cultivar expressing the Bt protein, it stops feeding and soon thereafter dies as a result of the Bt toxin binding to its gut wall. Bt corn is now commercially available in a number of countries to control corn borer (a lepidopteran insect), which is otherwise controlled by spraying (a more difficult process).

Crops have also been genetically engineered to acquire tolerance to broad-spectrum herbicide. The lack of herbicides with broad-spectrum activity and no crop injury was a consistent limitation in crop weed management. Multiple applications of numerous herbicides were routinely used to control a wide range of weed species detrimental to agronomic crops. Weed management tended to rely on preemergence—that is, herbicide applications were sprayed in response to expected weed infestations rather than in response to actual weeds present. Mechanical cultivation and hand weeding were often necessary to control weeds not controlled by herbicide applications.

The introduction of herbicide-tolerant crops has the potential of reducing the number of herbicide active ingredients used for weed management, reducing the number of herbicide applications made during a season, and increasing yield due to improved weed management and less crop injury. Transgenic crops that express tolerance to glyphosate, glufosinate and bromoxynil have been developed. These herbicides can now be sprayed on transgenic crops without inflicting damage on the crops while killing nearby weeds.

From 1996 to 2001, herbicide tolerance was the most dominant trait introduced to commercially available transgenic crops, followed by insect resistance. In 2001, herbicide tolerance deployed in soybean, corn and cotton accounted for 77% of the 626,000 square kilometres planted to transgenic crops; Bt crops accounted for 15%; and "stacked genes" for herbicide tolerance and insect resistance used in both cotton and corn accounted for 8%.

Production of Novel Substances in Crop Plants

Biotechnology is being applied for novel uses other than food. For example, oilseed can be modified to produce fatty acids for detergents, substitute fuels and petrochemicals. Potatoes, tomatoes, rice tobacco, lettuce, safflowers, and other plants have been genetically engineered to produce insulin and certain vaccines. If future clinical trials prove successful, the advantages of edible vaccines would be enormous, especially for developing countries.

The transgenic plants may be grown locally and cheaply. Homegrown vaccines would also avoid logistical and economic problems posed by having to transport traditional preparations over long distances and keeping them cold while in transit. And since they are edible, they will not need syringes, which are not only an additional expense in the traditional vaccine preparations but also a source of infections if contaminated.

In the case of insulin grown in transgenic plants, it is well-established that the gastrointestinal system breaks the protein down therefore this could not currently be administered as an edible protein. However, it might be produced at significantly lower cost than insulin produced in costly bioreactors. For example, Calgary, Canada-based SemBioSys Genetics, Inc. reports that its safflower-produced insulin will reduce unit costs by over 25% or more and approximates a reduction in the capital costs associated with building a commercial-scale insulin manufacturing facility of over $100 million, compared to traditional biomanufacturing facilities.

Criticism

There is another side to the agricultural biotechnology issue. It includes increased herbicide usage and resultant herbicide resistance, "super weeds," residues on and in food crops, genetic contamination of non-GM crops which hurt organic and conventional farmers, etc.

Biological Engineering

Biological engineering, biotechnological engineering or bioengineering (including biological systems engineering) is the application of concepts and methods of physics and mathematics to solve problems in life sciences, using engineering's own analytical and synthetical methodologies. In this context, while traditional engineering applies physical and mathematical sciences to analyse, design and manufacture inanimate tools, structures and processess, bioengineering uses the same sciences to study many aspects of living organisms. Usually it is used to analyse and solve problems related to human health. Biological engineering is a science based discipline founded upon the biological sciences in the same way that chemical engineering, electrical engineering, and mechanical engineering are based upon chemistry, electricity and magnetism, and statics, respectively.

Biological Engineering can be differentiated from its roots of pure biology or classical engineering in the following way. Biological studies often follow a reductionist approach in viewing a system on its smallest possible scale which naturally leads toward tools such as functional genomics. Engineering approaches, using classical design perspectives, are constructionist, building new devices, approaches, and technologies from component concepts. Biological engineering utilizes both of these methods in concert relying on reductionist approaches to define the fundamental units which are then commingled to generate something new. Although engineered biological systems have been used to manipulate information, construct materials, process chemicals, produce energy, provide food, and help maintain or enhance human health and our environment, our ability to quickly and reliably engineer biological systems that behave as expected remains less well developed than our mastery over mechanical and electrical systems.

The differentiation between Biological Engineering and overlap with Biomedical Engineering can be unclear, as many universities now use the terms "bioengineering" and "biomedical engineering" interchangeably. Some contend that Biological Engineering (like biotechnology) has a broader base which spans molecular methods (tends to emphasize the using of biological substances-applying engineering principles to molecular biology, biochemistry, microbiology, pharmacology, protein chemistry, cytology, immunology, neurobiology and neuroscience, cellular and tissue based methods (including devices and sensors), whole organisms (plants, animals), and up increasing length scales to ecosystems. Neither biological engineering nor biomedical engineering is wholly contained within the other, as there are non-biological products for medical needs and biological products for non-medical needs.

ABET, the U.S. based accreditation board for engineering B.S. programs, makes a distinction between Biomedical Engineering and Biological Engineering; however, the differences are quite small. Biomedical engineers must have life science courses that include human physiology and have experience in performing measurements on living systems while biological engineers must have life science courses (which may or may not include physiology) and experience in making measurements not specifically on living systems. Foundational engineering courses are often the same and include thermodynamics, fluid and mechanical dynamics, kinetics, electronics, and materials properties.

The word bioengineering was coined by British scientist and broadcaster Heinz Wolff in 1954. The term bioengineering is also used to describe the use of vegetation in civil engineering construction. The term bioengineering may also be applied to environmental modifications such as surface soil protection, slope stabilisation, watercourse and shoreline protection, windbreaks, vegetation barriers including noise barriers and visual screens, and the ecological enhancement of an area. The first biological engineering program was created at Mississippi State University in 1967, making it the first Biological Engineering curriculum in the United States. More recent programs have been launched at MIT and Utah State University.

Biological Engineers or *bioengineers* are engineers who use the principles of biology and the tools of engineering to create usable, tangible products. Biological Engineering employs knowledge and expertise from a number of pure and applied sciences, such as mass and heat transfer, kinetics, biocatalysts, biomechanics, bioinformatics, separation and purification processes, bioreactor design, surface science, fluid mechanics, thermodynamics, and polymer science. It is used in the design of medical devices, diagnostic equipment, biocompatible materials, renewable bioenergy, ecological engineering, and other areas that improve the living standards of societies.

In general, biological engineers attempt to either mimic biological systems in order to create products or modify and control biological systems so that they can replace, augment, or sustain chemical and mechanical processes. Bioengineers can apply their expertise to other applications of engineering and biotechnology, including genetic modification of plants and microorganisms, bioprocess engineering, and biocatalysis.

Because other engineering disciplines also address living organisms (e.g., prosthetics in mechanical engineering), the term biological engineering can be applied more broadly to include agricultural engineering and biotechnology. In fact, many old agricultural engineering departments in universities over the world have rebranded themselves as agricultural and biological engineering or agricultural and biosystems engineering. Biological engineering is also called bioengineering by some colleges and Biomedical engineering is called Bioengineering by others, and is a rapidly developing field with fluid categorization. The Main Fields of Bioengineering may be categorised as:

- Bioprocess Engineering: Bioprocess Design, Biocatalysis, Bioseparation, Bioinformatics, Bioenergy
- Genetic Engineering: Synthetic Biology, Horizontal gene transfer.
- Cellular Engineering: Cell Engineering, Tissue Culture Engineering, Metabolic Engineering.
- Biomedical Engineering: Biomedical technology, Biomedical Diagnostics, Biomedical Therapy, Biomechanics, Biomaterials.

Microbial Biodegradation

Interest in the microbial biodegradation of pollutants has intensified in recent years as humanity strives to find sustainable ways to cleanup contaminated environments. These bioremediation and biotransformation methods endeavour to harness the astonishing, naturally occurring, ability of microbial xenobiotic metabolism to degrade, transform or accumulate a huge range of compounds including hydrocarbons (e.g. oil), polychlorinated biphenyls (PCBs), polyaromatic hydrocarbons (PAHs), heterocyclic compounds (such as pyridine or quinoline), pharmaceutical substances, radionuclides and metals.

Major methodological breakthroughs in recent years have enabled detailed genomic, metagenomic, proteomic, bioinformatic and other high-throughput analyses of environmentally relevant microorganisms providing unprecedented insights into key biodegradative pathways and the ability of organisms to adapt to changing environmental conditions. The elimination of a wide range of pollutants and wastes from the environment is an absolute requirement to promote a sustainable development of our society with low environmental impact. Biological processes play a major role in the removal of contaminants and they take advantage of the astonishing catabolic versatility of microorganisms to degrade/convert such compounds. New methodological breakthroughs in sequencing, genomics, proteomics, bioinformatics and imaging are producing vast amounts of information.

In the field of Environmental Microbiology, genome-based global studies open a new era providing unprecedented *in silico* views of metabolic and regulatory networks, as well as clues to the evolution of degradation pathways and to the molecular adaptation strategies to changing environmental conditions. Functional genomic and metagenomic approaches are increasing our understanding of the relative importance of different pathways and regulatory networks to carbon flux in particular environments and for particular compounds and they will certainly accelerate the development of bioremediation technologies and biotransformation processes.

Aerobic Biodegradation of Pollutants

The burgeoning amount of bacterial genomic data provides unparalleled opportunities for understanding the genetic and molecular bases of the degradation of organic pollutants. Aromatic compounds are among the most recalcitrant of these pollutants and lessons can be learned from the recent genomic studies of *Burkholderia xenovorans* LB400 and *Rhodococcus* sp. strain RHA1, two of the largest bacterial genomes completely sequenced to date. These studies have helped expand our understanding of bacterial catabolism, non-catabolic physiological adaptation to organic compounds, and the evolution of large bacterial genomes. First, the metabolic pathways from phylogenetically diverse isolates are very similar with respect to overall organization.

Thus, as originally noted in pseudomonads, a large number of "peripheral aromatic" pathways funnel a range of natural and xenobiotic compounds into a restricted number of "central aromatic" pathways. Nevertheless, these pathways are genetically organized in genus-specific fashions, as exemplified by the b-ketoadipate and Paa pathways. Comparative genomic studies further reveal that some pathways are more widespread than initially thought.

Thus, the Box and Paa pathways illustrate the prevalence of non-oxygenolytic ring-cleavage strategies in aerobic aromatic degradation processes. Functional genomic studies have been useful in establishing that even organisms harboring high numbers of homologous enzymes seem to contain few examples of true redundancy. For example, the multiplicity of ring-cleaving dioxygenases in certain rhodococcal isolates may be attributed to the cryptic aromatic catabolism of different terpenoids and steroids.

Finally, analyses have indicated that recent genetic flux appears to have played a more significant role in the evolution of some large genomes, such as LB400's, than others. However, the emerging trend is that the large gene repertoires of potent pollutant degraders such as LB400 and RHA1 have evolved principally through more ancient processes. That this is true in such phylogenetically diverse species is remarkable and further suggests the ancient origin of this catabolic capacity.

Anaerobic Biodegradation of Pollutants

Anaerobic microbial mineralization of recalcitrant organic pollutants is of great environmental significance and involves intriguing novel biochemical reactions. In particular, hydrocarbons and halogenated compounds have long been doubted to be degradable in the absence of oxygen, but the isolation of hitherto unknown anaerobic hydrocarbon-degrading and reductively dehalogenating bacteria during the last decades provided ultimate proof for these processes in nature.

Many novel biochemical reactions were discovered enabling the respective metabolic pathways, but progress in the molecular understanding of these bacteria was rather slow, since genetic systems are not readily applicable for most of them. However, with the increasing application of genomics in the field of environmental microbiology, a new and promising perspective is now at hand to obtain molecular insights into these new metabolic properties. Several complete genome sequences were determined during the last few years from bacteria capable of anaerobic organic pollutant degradation. The ~4.7 Mb genome of the facultative denitrifying *Aromatoleum aromaticum* strain EbN1 was the first to be determined for an anaerobic hydrocarbon degrader (using toluene or ethylbenzene as substrates).

The genome sequence revealed about two dozen gene clusters (including several paralogs) coding for a complex catabolic network for anaerobic and aerobic degradation of aromatic compounds. The genome sequence forms the basis for current detailed studies on regulation of pathways and enzyme structures. Further genomes of anaerobic hydrocarbon degrading bacteria were recently completed for the iron-reducing species *Geobacter metallireducens* (accession nr. NC_007517) and the perchlorate-reducing *Dechloromonas aromatica* (accession nr. NC_007298), but these are not yet evaluated in formal publications.

Complete genomes were also determined for bacteria capable of anaerobic degradation of halogenated hydrocarbons by halorespiration: the ~1.4 Mb genomes of *Dehalococcoides ethenogenes* strain 195 and *Dehalococcoides* sp. strain CBDB1 and the ~5.7 Mb genome of *Desulfitobacterium hafniense* strain Y51. Characteristic for all these bacteria is the presence of multiple paralogous genes for reductive dehalogenases, implicating a wider dehalogenating spectrum of the organisms than previously known. Moreover, genome sequences provided unprecedented insights into the evolution of reductive dehalogenation and differing strategies for niche adaptation.

Recently, it has become apparent that some organisms, including *Desulfitobacterium chlororespirans*, originally evaluated for halorespiration on chlorophenols, can also use certain brominated compounds, such as the herbicide bromoxynil and its major metabolite as electron acceptors for growth. Iodinated compounds may be dehalogenated as well, though the process may not satisfy the need for an electron acceptor.

Bioavailability, Chemotaxis, and Transport of Pollutants

Bioavailability, or the amount of a substance that is physiochemically accessible to microorganisms is a key factor in the efficient biodegradation of pollutants. O'Loughlin et al (2000) showed that, with the exception of kaolinite clay, most soil clays and cation exchange resins attenuated biodegradation of 2-picoline by *Arthrobacter* sp. strain R1, as a result of adsorption of the substrate to the clays. Chemotaxis, or the directed movement of motile organisms towards or away from chemicals in the environment is an important physiological response that may contribute to effective catabolism of molecules in the environment. In addition, mechanisms for the intracellular accumulation of aromatic molecules via various transport mechanisms are also important.

Oil Biodegradation

Petroleum oil contains aromatic compounds that are toxic for most life forms. Episodic and chronic pollution of the environment by oil causes major ecological perturbations. Marine environments are especially vulnerable since oil spills of coastal regions and the open sea are poorly containable and mitigation is difficult. In addition to pollution through human activities, about 250 million litres of petroleum enter the marine environment every year from natural seepages. Despite its toxicity, a considerable fraction of petroleum oil entering marine systems is eliminated by the hydrocarbon-degrading activities of microbial communities, in particular by a remarkable recently discovered group of specialists, the so-called hydrocarbonoclastic bacteria (HCB). *Alcanivorax borkumensis* was the first HCB to have its genome sequenced. In addition to hydrocarbons, crude oil often contains various heterocyclic compounds, such as pyridine, which appear to be degraded by similar, though separate mechanisms than hydrocarbons.

Analysis of Waste Biotreatment

Sustainable development requires the promotion of environmental management and a constant search for new technologies to treat vast quantities of wastes generated by increasing anthropogenic activities. Biotreatment, the processing of wastes using living organisms, is an environmentally friendly, relatively simple and cost-effective alternative to physico-chemical clean-up options. Confined environments, such as bioreactors, have been engineered to overcome the physical, chemical and biological limiting factors of biotreatment processes in highly controlled systems. The great versatility in the design of confined environments allows the treatment of a wide range of wastes under optimized conditions. To perform a correct assessment, it is necessary to consider various microorganisms having a variety of genomes and expressed transcripts and proteins. A great number of analyses are often required. Using traditional genomic techniques, such assessments are limited and time-consuming. However, several high-throughput techniques originally developed for medical studies can be applied to assess biotreatment in confined environments.

Metabolic Engineering and Biocatalytic Applications

The study of the fate of persistent organic chemicals in the environment has revealed a large reservoir of enzymatic reactions with a large potential in preparative organic synthesis, which has already been exploited for a number of oxygenases on pilot and even on industrial scale. Novel catalysts can be obtained from metagenomic libraries and DNA sequence based approaches.

Our increasing capabilities in adapting the catalysts to specific reactions and process requirements by rational and random mutagenesis broadens the scope for application in the fine chemical industry, but also in the field of biodegradation. In many cases, these catalysts need to be exploited in whole cell bioconversions or in fermentations, calling for system-wide approaches to understanding strain physiology and metabolism and rational approaches to the engineering of whole cells as they are increasingly put forward in the area of systems biotechnology and synthetic biology.

Fungal Biodegradation

In the ecosystem, different substrates are attacked at different rates by consortia of organisms from different kingdoms. *Aspergillus* and other moulds play an important role in these consortia because they are adept at recycling starches, hemicelluloses, celluloses, pectins and other sugar polymers. Some aspergilli are capable of degrading more refractory compounds such as fats, oils, chitin, and keratin. Maximum decomposition occurs when there is sufficient nitrogen, phosphorus and other essential inorganic nutrients. Fungi also provide food for many soil organisms.

For *Aspergillus* the process of degradation is the means of obtaining nutrients. When these moulds degrade human-made substrates, the process usually is called biodeterioration. Both paper and textiles (cotton, jute, and linen) are particularly vulnerable to *Aspergillus* degradation. Our artistic heritage is also subject to *Aspergillus* assault. To give but one example, after Florence in Italy flooded in 1969, 74% of the isolates from a damaged Ghirlandaio fresco in the Ognissanti church were *Aspergillus versicolor*.

Biotechnology Regulations

The National Institute of Health was the first federal agency to assume regulatory responsibility in the United States. The Recombinant DNA Advisory Committee of the NIH published guidelines for working with recombinant DNA and recombinant organisms in the laboratory. Nowadays, the agencies that are responsible for the biotechnology regulation are: US Department of Agriculture (USDA) that regulates plant pests and medical preparation from living organisms, Environmental Protection Agency (EPA) that regulates pesticides and herbicides, and the Food and Drug Administration (FDA) which ensures that the food and drug products are safe and effective

Education

In 1988, after prompting from the United States Congress, the National Institute of General Medical Sciences (National Institutes of Health) instituted a funding mechanism for biotechnology training. Universities nationwide compete for these funds to establish Biotechnology Training Programs (BTPs). Each successful application is generally funded for five years then must be competitively renewed. Graduate students in turn compete for acceptance into a BTP; if accepted then stipend, tuition and health insurance support is provided for two or three years during the course of their PhD thesis work. Nineteen institutions offer NIGMS supported BTPs. Biotechnology training is also offered at the undergraduate level and in community colleges.

4

Molecular Phylogenetics

Molecular phylogenetics, also known as molecular systematics (a term likely discouraged to avoid confusion with molecular-biological system/structure-activity relationship), is the use of the structure of molecules to gain information on an organism's evolutionary relationships. The result of a molecular phylogenetic analysis is expressed in a phylogenetic tree.

History of Molecular Evolution

The history of molecular evolution starts in the early 20th century with "comparative biochemistry", but the field of molecular evolution came into its own in the 1960s and 1970s, following the rise of molecular biology. The advent of protein sequencing allowed molecular biologists to create phylogenies based sequence comparison, and to use the differences between homologous sequences as a molecular clock to estimate the time since the last common ancestor. In the late 1960s, the neutral theory of molecular evolution provided a theoretical basis for the molecular clock, though both the clock and the neutral theory were controversial, since most evolutionary biologists held strongly to panselectionism, with natural selection as the only important cause of evolutionary change. After the 1970s, nucleic acid sequencing allowed molecular evolution to reach beyond proteins to highly conserved ribosomal RNA sequences, the foundation of a reconceptualization of the early history of life.

Early History

Before the rise of molecular biology in the 1950s and 1960s, a small number of biologists had explored the possibilities of using biochemical differences between species to study evolution. Ernest Baldwin worked extensively on comparative biochemistry beginning in the 1930s, and Marcel Florkin pioneered techniques for constructing phylogenies based on molecular and biochemical characters in the 1940s. However, it was not until the 1950s that biologists developed techniques for producing biochemical data for the quantitative study of molecular evolution.

The first molecular systematics research was based on immunological assays and protein "fingerprinting" methods. Alan Boyden—building on immunological methods of G. H. F. Nuttall—developed new techniques beginning in 1954, and in the early 1960s Curtis Williams and Morris Goodman used immunological comparisons to study primate phylogeny. Others, such as Linus Pauling and his students, applied newly developed combinations of electrophoresis and paper chromatography to proteins subject to partial digestion by digestive enzymes to create unique two-dimensional patterns, allowing fine-grained comparisons of homologous proteins. Beginning in the 1950s, a few naturalists also experimented with molecular approaches—notably Ernst Mayr and Charles Sibley. While Mayr quickly soured on paper chromatography, Sibley successfully applied electrophoresis to egg-white proteins to sort out problems in bird taxonomy, soon supplemented that with DNA hybridization techniques— the beginning of a long career built on molecular systematics.

While such early biochemical techniques found grudging acceptance in the evolutionary biology community, for the most part they did not impact the main theoretical problems of evolution and population genetics. This would change as molecular biology shed more light on the physical and chemical nature of genes.

Genetic Load, the Classical/Balance Controversy, and the Measurement of Heterozygosity

At the time that molecular biology was coming into its own in the 1950s, there was a long-running debate—the classical/balance controversy—over the causes of heterosis, the increase in fitness observed when inbred lines are crossed. In 1950, James F. Crow offered two different explanations (later dubbed the *classical* and *balance* positions) based the paradox first articulated by J. B. S. Haldane in 1937: the effect of deleterious mutations on the average fitness of a population depends only on the rate of mutations (not the degree of harm caused by each mutation) because more-harmful mutations are eliminated more quickly by natural selection, while less-harmful mutations remain in the population longer. H. J. Muller dubbed this "genetic load".

Muller, motivated by his concern about the effects of radiation on human populations, argued that heterosis is primarily the result of deleterious homozygous recessive alleles, the effects of which are masked when separate lines are crossed—this was the *dominance hypothesis*, part of what Dobzhansky labelled the *classical position*. Thus, ionizing radiation and the resulting mutations produce considerable genetic load even if death or disease does not occur in the exposed generation, and in the absence of mutation natural selection will gradually increase the level of homozygosity. Bruce Wallace, working with J. C. King, used the *overdominance hypothesis* to develop the *balance position*, which left a larger place for overdominance (where the heterozygous state of a gene is more fit than the homozygous states). In that case, heterosis is simply the result of the increased expression of heterozygote advantage. If overdominant loci are common, then a high level of heterozygosity would result from natural selection, and mutation-induced radiation may in fact facilitate an increase in fitness due to overdominance. (This was also the view of Dobzhansky.)

Debate continued through 1950s, gradually becoming a central focus of population genetics. A 1958 study of *Drosophila* by Wallace suggested that radiation-induced mutations *increased* the viability of previously homozygous flies, providing evidence for heterozygote advantage and the balance position; Wallace estimated that 50% of loci in natural *Drosophila* populations were heterozygous. Motoo Kimura's subsequent mathematical analyses reinforced what Crow had suggested in 1950: that even if overdominant loci are rare, they could be responsible for a disproportionate amount of genetic variability. Accordingly, Kimura and his mentor Crow came down on the side of the classical position. Further collaboration between Crow and Kimura led to the infinite alleles model, which could be used to calculate the number of different alleles expected in a population, based on population size, mutation rate, and whether the mutant alleles were neutral, overdominant, or deleterious. Thus, the infinite alleles model offered a potential way to decide between the classical and balance positions, if accurate values for the level of heterozygosity could be found.

By the mid-1960s, the techniques of biochemistry and molecular biology—in particular, electrophoresis—provided a way to measure the level of heterozygosity in natural populations: a possible means to resolve the classical/balance controversy. In 1963, Jack L. Hubby published an electrophoresis study of protein variation in *Drosophila*; soon after, Hubby began collaborating with Richard Lewontin to apply Hubby's method to the classical/balance controversy by measuring the proportion of heterozygous loci in natural populations. Their two landmark papers, published in 1966, established a significant level of heterozygosity for *Drosophila* (12%, on average). However, these findings proved difficult to interpret. Most population geneticists (including Hubby and Lewontin) rejected the possibility of widespread neutral mutations; explanations that did not involve selection were anathema to mainstream evolutionary biology. Hubby and Lewontin also ruled out heterozygote advantage as the main cause because of the segregation load it would entail, though critics argued that the findings actually fit well with overdominance hypothesis.

Protein Sequences and the Molecular Clock

While evolutionary biologists were tentatively branching out into molecular biology, molecular biologists were rapidly turning their attention toward evolution.

After developing the fundamentals of protein sequencing with insulin between 1951 and 1955, Frederick Sanger and his colleagues had published a limited interspecies comparison of the insulin sequence in 1956. Francis Crick, Charles Sibley and others recognized the potential for using biological sequences to construct phylogenies, though few such sequences were yet available. By the early 1960s, techniques for protein sequencing had advanced to the point that direct comparison of homologous amino acid sequences was feasible. In 1961, Emanuel Margoliash and his collaborators completed the sequence for horse cytochrome c (a longer and more widely distributed protein than insulin), followed in short order by a number of other species.

In 1962, Linus Pauling and Emile Zuckerkandl proposed using the number of differences between homologous protein sequences to estimate the time since divergence, an idea Zuckerkandl had conceived around 1960 or 1961. This began with Pauling's long-time research focus, hemoglobin, which was being sequenced by Walter Schroeder; the sequences not only supported the accepted vertebrate phylogeny, but also the hypothesis (first proposed in 1957) that the different globin chains within a single organism could also be traced to a common ancestral protein. Between 1962 and 1965, Pauling and Zuckerkandl refined and elaborated this idea, which they dubbed the molecular clock, and Emil L. Smith and Emanuel Margoliash expanded the analysis to cytochrome c. Early molecular clock calculations agreed fairly well with established divergence times based on paleontological evidence. However, the essential idea of the molecular clock—that individual proteins evolve at a regular rate independent of a species' morphological evolution—was extremely provocative (as Pauling and Zuckerkandl intended it to be).

The Molecular Wars*f*

From the early 1960s, molecular biology was increasingly seen as a threat to the traditional core of evolutionary biology. Established evolutionary biologists—particularly Ernst Mayr, Theodosius Dobzhansky and G. G. Simpson, three of the founders of the modern evolutionary synthesis of the 1930s and 1940s—were extremely skeptical of molecular approaches, especially when it came to the connection (or lack thereof) to natural selection. Molecular evolution in general—and the molecular clock in particular—offered little basis for exploring evolutionary causation. According to the molecular clock hypothesis, proteins evolved essentially independently of the environmentally determined forces of selection; this was sharply at odds with the panselectionism prevalent at the time. Moreover, Pauling, Zuckerkandl, and other molecular biologists were increasingly bold in asserting the significance of "informational macromolecules" (DNA, RNA and proteins) for *all* biological processes, including evolution. The struggle between evolutionary biologists and molecular biologists— with each group holding up their discipline as the centre of biology as a whole—was later dubbed the "molecular wars" by Edward O. Wilson, who experienced firsthand the domination of his biology department by young molecular biologists in the late 1950s and the 1960s.

In 1961, Mayr began arguing for a clear distinction between *functional biology* (which considered proximate causes and asked "how" questions) and *evolutionary biology* (which considered ultimate causes and asked "why" questions) He argued that both disciplines and individual scientists could be classified on either the *functional* or *evolutionary* side, and that the two approaches to biology were complementary. Mayr, Dobzhansky, Simpson and others used this distinction to argue for the continued relevance of organismal biology, which was rapidly losing ground to molecular biology and related disciplines in the competition for funding and university support. It was in that context that Dobzhansky first published his famous statement, "nothing in biology makes sense except in the light of evolution", in a 1964 paper affirming the importance of organismal biology in the face of the molecular threat; Dobzhansky characterized the molecular disciplines as "Cartesian" (reductionist) and organismal disciplines as "Darwinian".

Mayr and Simpson attended many of the early conferences where molecular evolution was discussed, critiquing what they saw as the overly simplistic approaches of the molecular clock. The molecular clock, based on uniform rates of genetic change driven by random mutations and drift, seemed incompatible with the varying rates of evolution and environmentally-driven adaptive processes (such as adaptive radiation) that were among the key developments of the evolutionary synthesis. At the 1962 Wenner-Gren conference, the 1964 Colloquium on the Evolution of Blood Proteins in Bruges, Belgium, and the 1964 Conference on Evolving Genes and Proteins at Rutgers University, they engaged directly with the molecular biologists and biochemists, hoping to maintain the central place of Darwinian explanations in evolution as its study spread to new fields.

Gene-centered View of Evolution

Though not directly related to molecular evolution, the mid-1960s also saw the rise of the gene-centered view of evolution, spurred by George C. Williams's *Adaptation and Natural Selection* (1966). Debate over units of selection, particularly the controversy over group selection, led to increased focus on individual genes (rather than whole organisms or populations) as the theoretical basis for evolution. However, the increased focus on genes did not mean a focus on molecular evolution; in fact, the adaptationism promoted by Williams and other evolutionary theories further marginalized the apparently non-adaptive changes studied by molecular evolutionists.

The Neutral Theory of Molecular Evolution

The intellectual threat of molecular evolution became more explicit in 1968, when Motoo Kimura introduced the neutral theory of molecular evolution. Based on the available molecular clock studies (of hemoglobin from a wide variety of mammals, cytochrome c from mammals and birds, and triosephosphate dehydrogenase from rabbits and cows), Kimura (assisted by Tomoko Ohta) calculated an average rate of DNA substitution of one base pair change per 300 base pairs (encoding 100 amino acids) per 28 million years. For mammal genomes, this indicated a substitution rate of one every 1.8 years, which would produce an unsustainably high genetic load unless the preponderance of substitutions was selectively neutral. Kimura argued that neutral mutations occur very frequently, a conclusion compatible with the results of the electrophoretic studies of protein heterozygosity. Kimura also applied his earlier mathematical work on genetic drift to explain how neutral mutations could come to fixation, even in the absence of natural selection; he soon convinced James F. Crow of the potential power of neutral alleles and genetic drift as well.

Kimura's theory—described only briefly in a letter to *Nature*—was followed shortly after with a more substantial analysis by Jack L. King and Thomas H. Jukes—who titled their first paper on the subject "non-Darwinian evolution". Though King and Jukes produced much lower estimates of substitution rates and the resulting genetic load in the case of non-neutral changes, they agreed that neutral mutations driven by genetic drift were both real and significant.

The fairly constant rates of evolution observed for individual proteins was not easily explained without invoking neutral substitutions (though G. G. Simpson and Emil Smith had tried). Jukes and King also found a strong correlation between the frequency of amino acids and the number different of codons for each; this pointed to amino acid sequences as largely the product of random genetic drift.

King and Jukes' paper, especially with the provocative title, was seen as a direct challenge to mainstream neo-Darwinism, and it brought molecular evolution and the neutral theory to the centre of evolutionary biology. It provided a mechanism for the molecular clock and a theoretical basis for exploring deeper issues of molecular evolution, such as the relationship between rate of evolution and functional importance. The rise of the neutral theory marked synthesis of evolutionary biology and molecular biology—though an incomplete one. With their work on firmer theoretical footing, in 1971 Emile Zuckerkandl and other molecular evolutionists founded the *Journal of Molecular Evolution*.

The Neutralist-selectionist Debate and Near-neutrality

The critical responses to the neutral theory that soon appeared marked the beginning of the *neutralist-selectionist debate*. In short, selectionists viewed natural selection as the primary or only cause of evolution, even at the molecular level, while neutralists held that neutral mutations were widespread and that genetic drift was a crucial factor in the evolution of proteins. Kimura became the most prominent defender of the neutral theory—which would be his main focus for the rest of his career. With Ohta, he refocused his arguments on the rate at which drift could fix new mutations in finite populations, the significance of constant protein evolution rates, and the functional constraints on protein evolution that biochemists and molecular biologists had described. Though Kimura had initially developed the neutral theory partly as an outgrowth of the *classical position* within the classical/balance controversy (predicting high genetic load as a consequence of non-neutral mutations), he gradually deemphasized his original argument that segregational load would be impossibly high without neutral mutations (which many selectionists, and even fellow neutralists King and Jukes, rejected).

From the 1970s through the early 1980s, both selectionists and neutralists could explain the observed high levels of heterozygosity in natural populations, by assuming different values for unknown parameters. Early in the debate, Kimura's student Tomoko Ohta focused on the interaction between natural selection and genetic drift, which was significant for mutations that were not strictly neutral, but nearly so. In such cases, selection would compete with drift: most slightly deleterious mutations would be eliminated by natural selection or chance; some would move to fixation through drift. The Behaviour of this type of mutation, described by an equation that combined the mathematics of the neutral theory with classical models, became the basis of Ohta's nearly neutral theory of molecular evolution.

In 1973, Ohta published a short letter in *Nature* suggesting that a wide variety of molecular evidence supported the theory that most mutation events at the molecular level are slightly deleterious rather than strictly neutral. Molecular evolutionists were finding that while rates of protein evolution (consistent with the molecular clock) were fairly independent of generation time, rates of noncoding DNA divergence were inversely proportional to generation time. Noting that population size is generally inversely proportional to generation time, Tomoko Ohta proposed that most amino acid substitutions are slightly deleterious while noncoding DNA substitutions are more neutral. In this case, the faster rate of neutral evolution in proteins expected in small populations (due to genetic drift) is offset by longer generation times (and vice versa), but in large populations with short generation times, noncoding DNA evolves faster while protein evolution is retarded by selection (which is more significant than drift for large populations).

Between then and the early 1990s, many studies of molecular evolution used a "shift model" in which the negative effect on the fitness of a population due to deleterious mutations shifts back to an original value when a mutation reaches fixation. In the early 1990s, Ohta developed a "fixed model" that included both beneficial and deleterious mutations, so that no artificial "shift" of overall population fitness was necessary. According to Ohta, however, the nearly neutral theory largely fell out of favour in the late 1980s, because the mathematically simpler neutral theory for the widespread molecular systematics research that flourished after the advent of rapid DNA sequencing. As more detailed systematics studies started to compare the evolution of genome regions subject to strong selection versus weaker selection in the 1990s, the nearly neutral theory and the interaction between selection and drift have once again become an important focus of research.

Microbial Phylogeny

While early work in molecular evolution focused on readily sequenced proteins and relatively recent evolutionary history, by the late 1960s some molecular biologists were pushing further toward the base of the tree of life by studying highly conserved nucleic acid sequences. Carl Woese, a molecular biologist whose earlier work was on the genetic code and its origin, began using small subunit ribosomal RNA to reclassify bacteria by genetic (rather than morphological) similarity. Work proceeded slowly at first, but accelerated as new sequencing methods were developed in the 1970s and 1980s.

By 1977, Woese and George Fox announced that some bacteria, such as methanogens, lacked the rRNA units that Woese's phylogenetic studies were based on; they argued that these organisms were actually distinct enough from conventional bacteria and the so-called higher organisms to form their own kingdom, which they called archaebacteria. Though controversial at first (and challenged again in the late 1990s), Woese's work became the basis of the modern three-domain system of Archaea, Bacteria, and Eukarya (replacing the five-domain system that had emerged in the 1960s).

Work on microbial phylogeny also brought molecular evolution closer to cell biology and origin of life research. The differences between archaea pointed to the importance of RNA in the early history of life. In his work with the genetic code, Woese had suggested RNA-based life had preceded the current forms of DNA-based life, as had several others before him—an idea that Walter Gilbert would later call the "RNA world".

In many cases, genomics research in the 1990s produced phylogenies contradicting the rRNA-based results, leading to the recognition of widespread lateral gene transfer across distinct taxa. Combined with the probable endosymbiotic origin of organelle-filled eukarya, this pointed to a far more complex picture of the origin and early history of life, one which might not be describable in the traditional terms of common ancestry.

Techniques and Applications

Every living organism contains DNA, RNA, and proteins. Closely related organisms generally have a high degree of agreement in the molecular structure of these substances, while the molecules of organisms distantly related usually show a pattern of dissimilarity. Conserved sequences, such as mitochondrial DNA, are expected to accumulate mutations over time, and assuming a constant rate of mutation provide a molecular clock for dating divergence. Molecular phylogeny uses such data to build a "relationship tree" that shows the probable evolution of various organisms. Not until recent decades, however, has it been possible to isolate and identify these molecular structures.

The most common approach is the comparison of homologous sequences for genes using sequence alignment techniques to identify similarity. Another application of molecular phylogeny is in DNA barcoding, where the species of an individual organism is identified using small sections of mitochondrial DNA. Another application of the techniques that make this possible can be seen in the very limited field of human genetics, such as the ever more popular use of genetic testing to determine a child's paternity, as well as the emergence of a new branch of criminal forensics focused on evidence known as genetic fingerprinting.

Theoretical Background

Early attempts at molecular systematics were also termed as chemotaxonomy and made use of proteins, enzymes, carbohydrates and other molecules which were separated and characterized using techniques such as chromatography. These have been largely replaced in recent times by DNA sequencing which produces the exact sequences of nucleotides or *bases* in either DNA or RNA segments extracted using different techniques. These are generally considered superior for evolutionary studies since the actions of evolution are ultimately reflected in the genetic sequences. At present it is still a long and expensive process to sequence the entire DNA of an organism (its genome), and this has been done for only a few species. However it is quite feasible to determine the sequence of a defined area of a particular chromosome. Typical molecular systematic analyses require the sequencing of around 1000 base pairs. At any location within such a sequence, the bases found in a given position may vary between organisms. The particular sequence found in a given organism is referred to as its haplotype. In principle, since there are four base types, with 1000 base pairs, we could have 4 distinct haplotypes. However, for organisms within a particular species or in a group of related species, it has been found empirically that only a minority of sites show any variation at all and most of the variations that are found are correlated, so that the number of distinct haplotypes that are found is relatively small.

In a molecular systematic analysis, the haplotypes are determined for a defined area of genetic material; ideally a substantial sample of individuals of the target species or other taxon are used however many current studies are based on single individuals. Haplotypes of individuals of closely related, but supposedly different, taxa are also determined. Finally, haplotypes from a smaller number of individuals from a definitely different taxon are determined: these are referred to as an *out group*. The base sequences for the haplotypes are then compared. In the simplest case, the difference between two haplotypes is assessed by counting the number of locations where they have different bases: this is referred to as the number of *substitutions* (other kinds of differences between haplotypes can also occur, for example the *insertion* of a section of nucleic acid in one haplotype that is not present in another). Usually the difference between organisms is re-expressed as a *percentage divergence*, by dividing the number of substitutions by the number of base pairs analysed: the hope is that this measure will be independent of the location and length of the section of DNA that is sequenced.

An older and superseded approach was to determine the divergences between the genotypes of individuals by DNA-DNA hybridisation. The advantage claimed for using hybridisation rather than gene sequencing was that it was based on the entire genotype, rather than on particular sections of DNA. Modern sequence comparison techniques overcome this objection by the use of multiple sequences.

Once the divergences between all pairs of samples have been determined, the resulting triangular matrix of differences is submitted to some form of statistical cluster analysis, and the resulting dendrogram is examined in order to see whether the samples cluster in the way that would be expected from current ideas about the taxonomy of the group, or not. Any group of haplotypes that are all more similar to one another than any of them is to any other haplotype may be said to constitute a clade. Statistical techniques such as bootstrapping and jackknifing help in providing reliability estimates for the positions of haplotypes within the evolutionary trees.

Limitations of Molecular Systematics

Molecular systematics is an essentially cladistic approach: it assumes that classification must correspond to phylogenetic descent, and that all valid taxa must be monophyletic.

Molecular systematics often uses the molecular clock assumption that quantitative similarity of genotype is a sufficient measure of the recency of genetic divergence. Particularly in relation to speciation, this assumption could be wrong if either some genotypic modification acted to prevent interbreeding between two groups of organisms, or genetic modification proceeded at different rates in different subgroups of the organisms. In animals, it is often convenient to use mitochondrial DNA for molecular systematic analysis. However, because in mammals mitochondria are inherited only from the mother, this is not fully satisfactory, because inheritance in the paternal line might not be detected.

Molecular Medicine

Molecular medicine is a broad field, where physical, chemical, biological and medical techniques are used to describe molecular structures and mechanisms, identify fundamental molecular and genetic errors of disease, and to develop molecular interventions to correct them. The molecular medicine perspective emphasizes cellular and molecular phenomena and interventions rather than the previous conceptual and observational focus on patients and their organs.

In November, 1949, with the seminal paper, "Sickle Cell Anemia, a Molecular Disease", in *Science* magazine, Linus Pauling, Harvey Itano and their collaborators laid the groundwork for establishing the field of molecular medicine. In 1956, Roger J. Williams wrote *Biochemical Individuality*, a prescient book about genetics, prevention and treatment of disease on a molecular basis, and nutrition which is now variously referred to as individualized medicine and orthomolecular medicine. Another paper in *Science* by Pauling in 1968, introduced and defined this view of molecular medicine that focuses on natural and nutritional substances used for treatment and prevention. Published research and progress was slow until the 1970s' "biological revolution" that introduced many new techniques and commercial applications.

Molecular medicine is a new scientific discipline in European universities. Combining contemporary medical studies with the field of biochemistry, it offers a bridge between the two subjects. At present only a handful of universities offer the course to undergraduates. With a degree in this discipline the graduate is able to pursue a career in medical sciences, scientific research, laboratory work and postgraduate medical degrees.

Subjects

Core subjects are similar to biochemistry courses and typically include gene expression, research methods, proteins, cancer research, immunology, biotechnology and many more besides. In some universities molecular medicine is combined with another discipline such as chemistry, functioning as an additional study to enrich the undergraduate program.

Protein Structure

Proteins are an important class of biological macromolecules present in all organisms. All proteins are polymers of amino acids. Classified by their physical size, proteins are nanoparticles (definition: 1–100 nm). Each protein polymer – also known as a polypeptide – consists of a sequence of 20 different L-α-amino acids, also referred to as residues. For chains under 40 residues the term peptide is frequently used instead of protein. To be able to perform their biological function, proteins fold into one or more specific spatial conformations, driven by a number of non-covalent interactions such as hydrogen bonding, ionic interactions, Van Der Waals forces, and hydrophobic packing. To understand the functions of proteins at a molecular level, it is often necessary to determine their three-dimensional structure. This is the topic of the scientific field of structural biology, which employs techniques such as X-ray crystallography, NMR spectroscopy, and dual polarisation interferometry to determine the structure of proteins.

Protein structures range in size from tens to several thousand residues Very large aggregates can be formed from protein subunits: for example, many thousand actin molecules assemble into a microfilament. A protein may undergo reversible structural changes in performing its biological function. The alternative structures of the same protein are referred to as different conformations, and transitions between them are called conformational changes.

Protein Covalent Structure and Stereochemistry

Protein amino acids are combined into a single polypeptide chain in a condensation reaction. This reaction is catalysed by the ribosome in a process known as translation.

Amino Acid

Amino acids are molecules containing an amine group, a carboxylic acid group and a side chain that varies between different amino acids. The key elements of an amino acid are carbon, hydrogen, oxygen, and nitrogen. They are particularly important in biochemistry, where the term usually refers to *alpha-amino acids*.

An alpha-amino acid has the generic formula $H_2NCHRCOOH$, where R is an organic substituent; the amino group is attached to the carbon atom immediately adjacent to the carboxylate group (the α–carbon). Other types of amino acid exist when the amino group is attached to a different carbon atom; for example, in gamma-amino acids (such as gamma-amino-butyric acid) the carbon atom to which the amino group attaches is separated from the carboxylate group by two other carbon atoms. The various alpha-amino acids differ in which side chain (R-group) is attached to their alpha carbon, and can vary in size from just one hydrogen atom in glycine to a large heterocyclic group in tryptophan. Amino acids are critical to life, and have many functions in metabolism. One particularly important function is to serve as the building blocks of proteins, which are simply linear chains of amino acids. Just as the letters of the alphabet can be combined to form an almost endless variety of words, amino acids can be linked together in varying sequences to form a vast variety of proteins.

Due to their central role in biochemistry, amino acids are important in nutrition and are commonly used in food technology and industry. For example, monosodium glutamate is a common flavor enhancer that gives food the taste *umami*. In industry, applications include the production of biodegradable plastics, drugs, and chiral catalysts.

History

The first few amino acids were discovered in the early 19th century. In 1806, the French chemists Louis-Nicolas Vauquelin and Pierre Jean Robiquet isolated a compound in asparagus that proved to be asparagine, the first amino acid to be discovered. Another amino acid that was discovered in the early 19th century was cystine, in 1810, although its monomer, cysteine, was discovered much later, in 1884. Glycine and leucine were also discovered around this time, in 1820. Usage of the term *amino acid* in the English language is from 1898.

Proteinogenic Amino Acid

Proteinogenic amino acids are those amino acids that can be found in proteins and require cellular machinery coded for in the genetic code of any organism for their isolated production. There are 22 standard amino acids, but only 21 are found in eukaryotes. Of the twenty-two, twenty are directly encoded by the universal genetic code. Humans can synthesize 11 of these 20 from each other or from other molecules of intermediary metabolism, but the other 9 *essential amino acids* (histidine, isoleucine, leucine, lysine, methionine, phenylalanine, threonine, tryptophan, and valine) must be consumed in the diet. The remaining two, selenocysteine and pyrrolysine, are incorporated into proteins by unique synthetic mechanisms. *Proteinogenic* literally means *protein building*. Proteinogenic amino acids can be assembled into a polypeptide (the subunit of a protein) through a process known as translation (the second stage of protein biosynthesis, part of the overall process of gene expression).

Non-proteinogenic amino acids are either not found in proteins (like carnitine, GABA, or L-DOPA), or are not produced directly and in isolation by standard cellular machinery (like hydroxyproline and selenomethionine). The latter often results from posttranslational modification of proteins. There are clear reasons why organisms have not evolved to incorporate certain non-proteinogenic amino acids into proteins: for example, ornithine and homoserine will cyclize against the peptide backbone and fragment the protein with relatively short half-lives, and others are toxic because they can be mistakenly incorporated into proteins, such as the arginine analog canavanine.

Non-proteinogenic amino acids are found in nonribosomal peptides, which are not produced by the ribosome during translation.

Structures

The following illustrates the structures and abbreviations of the 21 amino acids that are directly encoded for protein synthesis by the genetic code of eukaryotes. The structures given below are standard chemical structures, not the typical zwitterion forms that exist in aqueous solutions.

Non-specific Abbreviations

Sometimes the specific identity of an amino acid cannot be determined unambiguously. Certain protein sequencing techniques do not distinguish among certain pairs. Thus, the following codes are used:

- *Asx* (B) is "asparagine or aspartic acid"
- *Glx* (Z) is "glutamic acid or glutamine"
- *Xle* (J) is "leucine or isoleucine".

In addition, the symbol X is used to indicate an amino acid that is completely unidentified.

Chemical Properties

Following is a table listing the one-letter symbols, the three-letter symbols, and the chemical properties of the side-chains of the standard amino acids. The masses listed are based on weighted averages of the elemental isotopes at their natural abundances. Note that forming a peptide bond results in elimination of a molecule of water, so the mass of an amino acid unit within a protein chain is reduced by 18.01524 Da.

Table 1: General Chemical Properties

Amino Acid	Short	Abbrev.	Avg. Mass (Da)	pI	pK₁(α-COOH)	pK₂(α-⁺NH₃)
Alanine	A	Ala	89.09404	6.01	2.35	9.87
Cysteine	C	Cys	121.15404	5.05	1.92	10.70
Aspartic acid	D	Asp	133.10384	2.85	1.99	9.90
Glutamic acid	E	Glu	147.13074	3.15	2.10	9.47
Phenylalanine	F	Phe	165.19184	5.49	2.20	9.31
Glycine	G	Gly	75.06714	6.06	2.35	9.78
Histidine	H	His	155.15634	7.60	1.80	9.33
Isoleucine	I	Ile	131.17464	6.05	2.32	9.76
Lysine	K	Lys	146.18934	9.60	2.16	9.06
Leucine	L	Leu	131.17464	6.01	2.33	9.74
Methionine	M	Met	149.20784	5.74	2.13	9.28
Asparagine	N	Asn	132.11904	5.41	2.14	8.72
Pyrrolysine	O	Pyl				
Proline	P	Pro	115.13194	6.30	1.95	10.64
Glutamine	Q	Gln	146.14594	5.65	2.17	9.13
Arginine	R	Arg	174.20274	10.76	1.82	8.99
Serine	S	Ser	105.09344	5.68	2.19	9.21
Threonine	T	Thr	119.12034	5.60	2.09	9.10
Selenocysteine	U	Sec	168.053			
Valine	V	Val	117.14784	6.00	2.39	9.74
Tryptophan	W	Trp	204.22844	5.89	2.46	9.41
Tyrosine	Y	Tyr	181.19124	5.64	2.20	9.21

Isomerism

Of the standard α-amino acids, all but glycine can exist in either of two optical isomers, called L or D amino acids, which are mirror images of each other. While L-amino acids represent all of the amino acids found in proteins during translation in the ribosome, D-amino acids are found in some proteins produced by enzyme posttranslational modifications after translation and translocation to the endoplasmic reticulum, as in exotic sea-dwelling organisms such as cone snails. They are also abundant components of the peptidoglycan cell walls of bacteria, and D-serine may act as a neurotransmitter in the brain. The L and D convention for amino acid configuration refers not to the optical activity of the amino acid itself, but rather to the optical activity of the isomer of glyceraldehyde from which that amino acid can theoretically be synthesized (D-glyceraldehyde is dextrorotary; L-glyceraldehyde is levorotary).

Alternatively, the *(S)* and *(R)* designators are used to indicate the absolute stereochemistry. Almost all of the amino acids in proteins are *(S)* at the α carbon, with cysteine being *(R)* and glycine non-chiral. Cysteine is unusual since it has a sulfur atom at the second position in its side-chain, which has a larger atomic mass than the groups attached to the first carbon which is attached to the α-carbon in the other standard amino acids, thus the *(R)* instead of *(S)*.

Zwitterions

The amine and carboxylic acid functional groups found in amino acids allow it to have amphiprotic properties. At a certain pH, known as the isoelectric point, an amino acid has no overall charge since the number of protonated ammonia groups (positive charges) and deprotonated carboxylate groups (negative charges) are equal. The amino acids all have different isoelectric points. The ions produced at the isoelectric point have both positive and negative charges and are known as a *zwitterion,* which comes from the German word *Zwitter* meaning "hermaphrodite" or "hybrid". Amino acids can exist as zwitterions in solids and in polar solutions such as water, but not in the gas phase. Zwitterions have minimal solubility at their isolectric point and an amino acid can be isolated by precipitating it from water by adjusting the pH to its particular isoelectric point.

Protein Primary Structure

The primary structure of peptides and proteins refers to the linear sequence of its amino acid structural units. The term "primary structure" was first coined by Linderstrom-Lang in 1951. By convention, the primary structure of a protein is reported starting from the amino-terminal (N) end to the carboxyl-terminal (C) end.

Primary Structure of Polypeptides

In general, polypeptides are unbranched polymers, so their primary structure can often be specified by the sequence of amino acids along their backbone. However, proteins can become cross-linked, most commonly by disulfide bonds, and the primary structure also requires specifying the cross-linking atoms, e.g., specifying the cysteines involved in the protein's disulfide bonds. Other crosslinks include desmosine... The chiral centres of a polypeptide chain can undergo racemization. In particular, the L-amino acids normally found in proteins can spontaneously isomerize at the Cα atom to form D-amino acids, which cannot be cleaved by most proteases.

Finally, the protein can undergo a variety of posttranslational modifications, which are briefly summarized here. The N-terminal amino group of a polypeptide can be modified covalently, e.g.,

- acetylation – $C(= O) - CH_3$: The positive charge on the N-terminal amino group may be eliminated by changing it to an acetyl group (N-terminal blocking).

- formylation – $C(= O)H$: The N-terminal methionine usually found after translation has an N-terminus blocked with a formyl group. This formyl group (and sometimes the methionine residue itself, if followed by Gly or Ser) is removed by the enzyme deformylase.

- Pyroglutamate.

Formation of pyroglutamate from an N-terminal glutamine:

An N-terminal glutamine can attack itself, forming a cyclic pyroglutamate group.

- Myristoylation: Similar to acetylation. Instead of a simple methyl group, the myristoyl group has a tail of 14 hydrophobic carbons, which make it ideal for anchoring proteins to cellular membranes.

The C-terminal carboxylate group of a polypeptide can also be modified, e.g.,

C-terminal amidation it:

- Amidation: The C-terminus can also be blocked (thus, neutralizing its negative charge) by amidation.

- glycosyl phosphatidylinositol (GPI) attachment : Glycosyl phosphatidylinositol is a large, hydrophobic phospholipid prosthetic group that achors proteins to cellular membranes. It is attached to the polypeptide C-terminus through an amide linkage that then connects to ethanolamine, thence to sundry sugars and finally to the phosphatidylinositol lipid moiety.

Finally, the peptide side chains can also be modified covalently, e.g.,

- Phosphorylation: Aside from cleavage, phosphorylation is perhaps the most important chemical modification of proteins. A phosphate group can be attached to the sidechain hydroxyl group of serine, threonine and tyrosine residues, adding a negative charge at that site and producing an unnatural amino acid. Such reactions are catalysed by kinases and the reverse reaction is catalysed by phosphatases. The phosphorylated tyrosines are often used as "handles" by which proteins can bind to one another, whereas phosphorylation of Ser/Thr often induces conformational changes, presumably because of the introduced negative charge. The effects of phosphorylating Ser/Thr can sometimes be simulated by mutating the Ser/Thr residue to glutamate.

Glycosylation

Glycosylation is the enzymatic process that attaches glycans to proteins, lipids, or other organic molecules. This enzymatic process produces one of the fundamental biopolymers found in cells (along with DNA, RNA, and proteins). Glycosylation is a form of co-translational and post-translational modification. Glycans serve a variety of structural and functional roles in membrane and secreted proteins. The majority of proteins synthesized in the rough ER undergo glycosylation. It is an enzyme-directed site-specific process, as opposed to the non-enzymatic chemical reaction of glycation. Glycosylation is also present in the cytoplasm and nucleus as the O-GlcNAc modification. Five classes of glycans are produced:

- *N*-linked glycans attached to a nitrogen of asparagine or arginine side chains;
- *O*-linked glycans attached to the hydroxy oxygen of serine, threonine, tyrosine, hydroxylysine, or hydroxyproline side chains, or to oxygens on lipids such as ceramide;
- phospho-glycans linked through the phosphate of a phospho-serine;
- *C*-linked glycans, a rare form of glycosylation where a sugar is added to a carbon on a tryptophan side chain;
- glypiation, which is the addition of a GPI anchor that links proteins to lipids through glycan linkages.

Purpose

The carbohydrate chains attached to the target proteins serve various functions. For instance, some proteins do not fold correctly unless they are glycosylated first. Also, polysaccharides linked at the amide nitrogen of asparagine in the protein confer stability on some secreted glycoproteins. Experiments have shown that glycosylation in this case is not a strict requirement for proper folding, but the unglycosylated protein degrades quickly. Glycosylation may play a role in cell-cell adhesion (a mechanism employed by cells of the immune system), as well.

Mechanisms

There are various mechanisms for glycosylation, although most share several common features:

* Glycosylation, unlike glycation, is an enzymatic process;
* The donor molecule is often an activated nucleotide sugar;
* The process is site-specific.

Types of Glycosylation

N-linked Glycosylation: N-linked glycosylation is important for the folding of some eukaryotic proteins. The N-linked glycosylation process occurs in eukaryotes and widely in archaea, but very rarely in bacteria.

O-linked Glycosylation: O-linked glycosylation is a form of glycosylation occurring in the Golgi apparatus.

Phospho-Serine Glycosylation: Xylose, fucose, mannose, and GlcNAc phospho-serine glycans have been reported in the literature. Fucose and GlcNAc have been found only in *Dictyostelium discoideum*, mannose in *Leishmania mexicana*, and xylose in *Trypanosoma cruzi*.

C-mannosylation: A mannose sugar is added to the first tryptophan residue in the sequence W-X-X-W (W indicates tryptophan, X is any amino acid). Thrombospondins are one of the most commonly modified proteins, however this form of glycosylation appears elsewhere as well. This is an unusual modification because the sugar is linked to a carbon rather than a reactive atom like a nitrogen or oxygen.

GPI Anchors (Glypiation): A special form of glycosylation is the *GPI anchor*. This form of glycosylation functions to attach a protein to a hydrophobic lipid anchor, via a glycan chain.

Deamidation

Deamidation is a chemical reaction in which an amide functional group is removed from an organic compound. In biochemistry, the reaction is important in the degradation of proteins because it damages the amide-containing side chains of the amino acids asparagine and glutamine.

In the biochemical deamidation reaction, the side chain of an asparagine attacks the following peptide group (in black at top right of Figure), forming a symmetric succinimide intermediate (in red). The symmetry of the intermediate results in two products of its hydrolysis, either aspartate (in black at left) or in isoaspartate, which is a beta amino acid (in green at bottom right). This process is considered a deamidation because the amide in the asparagine side chain is replaced by a carboxylate group. However, a similar reaction can occur in aspartate side chains, yielding a partial conversion to isoaspartate.

Kinetics of Deamidation

Deamidation reactions have been conjectured to be one of the factors that limit the useful lifetime of proteins.

Deamidation proceeds much more quickly if the susceptible amino acid is followed by a small, flexible residue such as glycine whose low steric hindrance leaves the peptide group open for attack. Deamidation reactions also proceed much more quickly at elevated pH (>10) and temperature.

Hydroxylation

Hydroxylation is a chemical process that introduces a hydroxyl groups (-OH) into an organic compound. In biochemistry, hydroxylation reactions are often facilitated by enzymes called hydroxylases. Hydroxylation is the first step in the oxidative degradation of organic compounds in air. It is extremely important in detoxification since hydroxylation converts lipophilic compounds into water-soluble (hydrophilic) products that are more readily excreted. Some drugs (e.g. steroids) are activated or deactivated by hydroxylation.

Chemical Concepts

The hydroxylation process involves conversion of a CH group into a COH group. Hydroxylation is an oxidative process. The oxygen that is inserted into the C-H bond is usually derived from atmospheric oxygen (O_2). Since O_2 itself is a slow hydroxylating agent, catalysts are required to accelerate the pace of the process.

Biological Hydroxylation

The principal hydroxylation agent in nature is cytochrome P-450, hundreds of variations of which are known. Other hydroxylating agents include flavins.

Hydroxylation of Proteins

The principal residue to be hydroxylated in proteins is proline. The hydroxylation occurs at the γ-C atom, forming hydroxyproline

(Hyp), an essential element of collagen, in turn a necessary element of connective tissue. Proline hydroxylation is also a vital component of hypoxia response via hypoxia inducible factors. In some cases, proline may be hydroxylated instead on its β-C atom. Lysine may also be hydroxylated on its δ-C atom, forming hydroxylysine (Hyl).

These three reactions are catalysed by very large, multi-subunit enzymes prolyl 4-hydroxylase, prolyl 3-hydroxylase and lysyl 5-hydroxylase, respectively. These reactions require iron (as well as molecular oxygen and α-ketoglutarate) to carry out the oxidation, and use ascorbic acid (vitamin C) to return the iron to its oxidized state. Deprivation of ascorbate leads to deficiencies in proline hydroxylation, which leads to less stable collagen, which can manifest itself as the disease scurvy. Since citrus fruits are rich in vitamin C, British sailors were given limes to combat scurvy on long ocean voyages; hence, they were called "limeys".

Examples of Hydroxylases

- 17α-hydroxylase
- Cholesterol 7 alpha-hydroxylase
- Dopamine β-hydroxylase
- Phenylalanine hydroxylase
- Tyrosine hydroxylase.

Methylation

In the chemical sciences, methylation denotes the addition of a methyl group to a substrate or the substitution of an atom or group by a methyl group. Methylation is a form of alkylation with, to be specific, a methyl group, rather than a larger carbon chain, replacing a hydrogen atom. These terms are commonly used in chemistry, biochemistry, soil science, and the biological sciences.

In biological systems, methylation is catalysed by enzymes; such methylation can be involved in modification of heavy metals, regulation of gene expression, regulation of protein function, and RNA metabolism. Methylation of heavy metals can also occur outside of biological systems. Chemical methylation of tissue samples is also one method for reducing certain histological staining artifacts.

Biological Methylation

Epigenetics: Methylation contributing to epigenetic inheritance can occur through either DNA methylation or protein methylation.

DNA methylation in vertebrates typically occurs at CpG sites (cytosine-phosphate-guanine sites, that is, where a cytosine is directly followed by a guanine in the DNA sequence). This methylation results in the conversion of the cytosine to 5-methylcytosine. The formation of Me-CpG is catalysed by the enzyme DNA methyltransferase. Human DNA has about 80%-90% of CpG sites methylated, but there are certain areas, known as CpG islands, that are GC-rich (made up of about 65% CG residues), wherein none are methylated. These are associated with the promoters of 56% of mammalian genes, including all ubiquitously expressed genes. One to two percent of the human genome are CpG clusters, and there is an inverse relationship between CpG methylation and transcriptional activity.

Protein methylation typically takes place on arginine or lysine amino acid residues in the protein sequence. Arginine can be methylated once (monomethylated arginine) or twice, with either both methyl groups on one terminal nitrogen (asymmetric dimethylated arginine) or one on both nitrogens (symmetric dimethylated arginine) by peptidylarginine methyltransferases (PRMTs). Lysine can be methylated once, twice or three times by lysine methyltransferases. Protein methylation has been most-studied in the histones. The transfer of methyl groups from S-adenosyl methionine to histones is catalysed by enzymes known as histone methyltransferases. Histones that are methylated on certain residues can act epigenetically to repress or activate gene expression. Protein methylation is one type of post-translational modification.

Embryonic Development

During the development of germ cells their genomes are demethylated, while chromosomes in the somatic cells retain the parental methylation patterns. After that, a *De novo* methylation of the germ cells occurs, modifying and adding epigenetic information to the genome based on the sex of the individual. After fertilization of an oocyte and formations of a zigote, its combined genome is demethylated and remethylated again (with the exception of the imprinted genes). By blastula stage, the methylation of the embryonic cells is complete. The process of demethylation/remethylation is referred to as "reprogramming". The importance of methylation was shown in knockout mutants without DNA methyltransferase, which all died at the morula stage.

Postnatal Development

Increasing evidence is revealing a role of methylation in the interaction of environmental factors with genetic expression.

Differences in maternal care during the first 6 days of life in the rat induce differential methylation patterns in some promoter regions and, thus, influencing gene expression. Furthermore, even-more-dynamic processes such as interleukin signalling have been shown to be regulated by methylation.

Cancer

The pattern of methylation has recently become an important topic for research. Studies have found that in normal tissue, methylation of a gene is mainly localized to the coding region, which is CpG-poor. In contrast, the promoter region of the gene is unmethylated, despite a high density of CpG islands in the region.

Neoplasia is characterized by "methylation imbalance" where genome-wide hypomethylation is accompanied by localized hypermethylation and an increase in expression of DNA methyltransferase. The overall methylation state in a cell might also be a precipitating factor in carcinogenesis as evidence suggests that genome-wide hypomethylation can lead to chromosome instability and increased mutation rates. The methylation state of some genes can be used as a biomarker for tumorigenesis. For instance, hypermethylation of the pi-class glutathione S-transferase gene (GSTP1) appears to be a promising diagnostic indicator of prostate cancer. In cancer, the dynamics of genetic and epigenetic gene silencing are very different. Somatic genetic mutation leads to a block in the production of functional protein from the mutant allele. If a selective advantage is conferred to the cell, the cells expand clonally to give rise to a tumour in which all cells lack the capacity to produce protein. In contrast, epigenetically mediated gene silencing occurs gradually. It begins with a subtle decrease in transcription, fostering a decrease in protection of the CpG island from the spread of flanking heterochromatin and methylation into the island. This loss results in gradual increases of individual CpG sites, which vary between copies of the same gene in different cells.

Bacterial Host Defence

In addition, adenosine or cytosine methylation is part of the restriction modification system of many bacteria. Bacterial DNAs are methylated periodically throughout the genome. A methylase is the enzyme that recognizes a specific sequence and methylates one of the bases in or near that sequence. Foreign DNAs (which are not methylated in this manner) that are introduced into the cell are degraded by sequence-specific restriction enzymes. Bacterial genomic DNA is not recognized by these restriction enzymes. The methylation of native DNA acts as a sort of primitive immune system, allowing the bacteria to protect themselves from infection by bacteriophage. These restriction enzymes are the basis of restriction fragment length polymorphism (RFLP) testing, used to detect DNA polymorphisms.

Methylation in Chemistry

The term methylation in organic chemistry refers to the alkylation process used to describe the delivery of a CH_3 group. This is commonly performed using *electrophilic* methyl sources-iodomethane, dimethyl sulfate, dimethyl carbonate, or less commonly with the more powerful (and more dangerous) methylating reagents of methyl triflate or methyl fluorosulfonate (magic methyl), which all react via S_N2 nucleophilic substitution. For example a carboxylate may be methylated on oxygen to give a methyl ester, an alkoxide salt RO$^-$ may be likewise methylated to give an ether, $ROCH_3$, or a ketone enolate may be methylated on carbon to produce a new ketone. On the other hand, the methylation may involve use of *nucleophilic* methyl compounds such as methyllithium (CH_3Li) or Grignard reagents (CH_3MgX). For example, CH_3Li will methylate acetone, adding across the carbonyl (C=O) to give the lithium alkoxide of *tert*-butanol:

Purdie Methylation

Purdie methylation is a specific method for the methylation at oxygen of carbohydrates using iodomethane and silver oxide.

Acetylation

Acetylation (or in IUPAC nomenclature ethanoylation) describes a reaction that introduces an acetyl functional group into a chemical compound. Deacetylation is the removal of the acetyl group.

Moreover, it is that process of introducing an acetyl group (resulting in an acetoxy group) into a compound, to be specific, the substitution of an acetyl group for an active hydrogen atom. A reaction involving the replacement of the hydrogen atom of a hydroxyl group with an acetyl group (CH_3 CO) yields a specific ester, the acetate. Acetic anhydride is commonly used as an acetylating agent reacting with free hydroxyl groups. For example, it is used in the synthesis of aspirin and heroin.

Acetylation of Proteins

In biology, i.e., in living cells, acetylation occurs as a co-translational and post-translational modification of proteins, for example, histones, p53, and tubulins.

N-alpha-terminal Acetylation

Acetylation of the N-terminal alpha-amine of proteins is a widespread modification in eukaryotes. Forty to fifty percent of yeast proteins, and 80-90% of human proteins are modified in this manner, and the pattern of modification is found to be conserved throughout evolution. The modification is performed by N-alpha-acetyltransferases (NATs), a sub-family of the GNAT superfamily of acetyltransferases, which also include histone acetyl transferases. The GNATs transfer the acetylgroup from acetyl-coenzyme A to the amine group. The NATs have been most extensively studied in yeast. Here, three NAT complexes, NatA/B/C, have been found to perform most N-alpha-terminal acetylations. They have sequence specificity for their substrates, and it is believed that they are associated with the ribosome, where they acetylate the nascent polypeptide chain. In humans, the human NatA and NatB complexes have been identified and characterized. Subunits of the human NatA complex have been coupled to cancer-related processes such as hypoxia-response and the beta-catenin pathway. It has been found to be over-expressed in papillary thyroid carcinoma and neuroblastoma. The human NatB complex have been coupled to cell cycle. The hNat3 subunit of the hNatB complex has been found overexpressed in some forms of cancer.

Despite being a conserved and widespread modification, little is known about the biological role of N-alpha-terminal acetylation. Proteins such as actin and tropomyosin have been found to be dependent of NatB acetylation to form proper actin filaments. Yet this is only one example pointing to the potential importance of this modification.

For unknown reasons 20% of Asians have an isozyme that results in slower N-acetylation of drugs, while 50% of Whites and African-Americans do.

Lysine Acetylation and Deacetylation

In histone acetylation and deacetylation, the histones are acetylated and deacetylated on lysine residues in the N-terminal tail as part of gene regulation. Typically, these reactions are catalysed by enzymes with "histone acetyltransferase" (HAT) or "histone deacetylase" (HDAC) activity, although HATs and HDACs can modify the acetylation status of non-histone proteins as well. The regulation of transcription factors, effector proteins, molecular chaperones, and cytoskeletal proteins by acetylation/deacetylation is emerging as a significant post-translational regulatory mechanism analogous to phosphorylation by the action of kinases or dephosphorylated by the action of phosphatases. Not only can the acetylation state of a protein modify its activity, there has been recent suggestion that this post-translational modification might crosstalk with phosphorylation, methylation, ubiquitination, sumoylation, and others for dynamic control of cellular signalling.

The tubulin acetylation and deacetylation system is well worked out in Chlamydomonas. A tubulin acetyltransferase located in the axoneme acetylates a specific lysine residue in the α-tubulin subunit in assembled microtubule. Once disassembled, this acetylation can be removed by another specific deacetylase that is cytosolic. Thus the axonemal microtubules (long half-life) carry this signature acetylation absent from cytosolic microtubules (short half-life).

- sulfation : Tyrosines may become sulfated on their O^η atom. Somewhat unusually, this modification occurs in the Golgi apparatus, not in the endoplasmic reticulum. Similar to phosphorylated tyrosines, sulfated tyrosines are used for specific recognition, e.g., in chemokine receptors on the cell surface. As with phosphorylation, sulfation adds a negative charge to a previously neutral site.

- prenylation and palmitoylation : The hydrophobic isoprene (e.g., farnesyl, geranyl, and geranylgeranyl groups) and palmitoyl groups may be added to the S^γ atom of cysteine residues to anchor proteins to cellular membranes. Unlike the GPI and myritoyl anchors, these groups are not necessarily added at the termini.

- carboxylation : A relatively rare modification that adds an extra carboxylate group (and, hence, a double negative charge) to a glutamate side chain, producing a Gla residue. This is used to strengthen the binding to "hard" metal ions such as calcium.

- ADP-ribosylation : The large ADP-ribosyl group can be transferred to several types of side chains within proteins, with heterogeneous effects. This modification is a target for the powerful toxins of disparate bacteria, e.g., *Vibrio cholerae*, *Corynebacterium diphtheriae* and *Bordetella pertussis*.

- ubiquitination and SUMOylation : Various full-length, folded proteins can be attached at their C-termini to the sidechain ammonium groups of lysines of other proteins. Ubiquitin is the most common of these, and usually signals that the ubiquitin-tagged protein should be degraded.

Most of the polypeptide modifications listed above occur *post-translationally*, i.e., after the protein has been synthesized on the ribosome, typically occurring in the endoplasmic reticulum, a subcellular organelle of the eukaryotic cell.

Many other chemical reactions (e.g., cyanylation) have been applied to proteins by chemists, although they are not found in biological systems.

Modifications of Primary Structure

In addition to those listed above, the most important modification of primary structure is peptide cleavage. Proteins are often synthesized in an inactive precursor form; typically, an N-terminal or C-terminal segment blocks the active site of the protein, inhibiting its function. The protein is activated by cleaving off the inhibitory peptide.

Some proteins even have the power to cleave themselves. Typically, the hydroxyl group of a serine (rarely, threonine) or the thiol group of a cysteine residue will attack the carbonyl carbon of the preceding peptide bond, forming a tetrahedrally bonded intermediate [classified as a hydroxyoxazolidine (Ser/Thr) or hydroxythiazolidine (Cys) intermediate]. This intermediate tends to revert to the amide form, expelling the attacking group, since the amide form is usually favored by free energy, (presumably due to the strong resonance stabilization of the peptide group). However, additional molecular interactions may render the amide form less stable; the amino group is expelled instead, resulting in an ester (Ser/Thr) or thioester (Cys) bond in place of the peptide bond. This chemical reaction is called an N-O acyl shift.

The ester/thioester bond can be resolved in several ways:

- Simple hydrolysis will split the polypeptide chain, where the displaced amino group becomes the new N-terminus. This is seen in the maturation of glycosylasparaginase.

- A β-elimination reaction also splits the chain, but results in a pyruvoyl group at the new N-terminus. This pyruvoyl group may be used as a covalently attached catalytic cofactor in some enzymes, especially decarboxylases such as S-adenosylmethionine decarboxylase (SAMDC) that exploit the electron-withdrawing power of the pyruvoyl group.

- Intramolecular transesterification, resulting in a *branched* polypeptide. In inteins, the new ester bond is broken by an intramolecular attack by the soon-to-be C-terminal asparagine.

- Intermolecular transesterification can transfer a whole segment from one polypeptide to another, as is seen in the Hedgehog protein autoprocessing.

History of Protein Primary Structure

The proposal that proteins were linear chains of α-amino acids was made nearly simultaneously by two scientists at the same conference in 1902, the 74th meeting of the Society of German Scientists and Physicians, held in Karlsbad. Franz Hofmeister made the proposal in the morning, based on his observations of the biuret reaction in proteins. Hofmeister was followed a few hours later by Emil Fischer, who had amassed a wealth of chemical details supporting the peptide-bond model. For completeness, the proposal that proteins contained amide linkages was made as early as 1882 by the French chemist E. Grimaux.

Despite these data and later evidence that proteolytically digested proteins yielded only oligopeptides, the idea that proteins were linear, unbranched polymers of amino acids was not accepted immediately. Some well-respected scientists such as William Astbury doubted that covalent bonds were strong enough to hold such long molecules together; they feared that thermal agitations would shake such long molecules asunder.

Hermann Staudinger faced similar prejudices in the 1920s when he argued that rubber was composed of macromolecules.

Thus, several alternative hypotheses arose. The colloidal protein hypothesis stated that proteins were colloidal assemblies of smaller molecules. This hypothesis was disproved in the 1920s by ultracentrifugation measurements by Theodor Svedberg that showed that proteins had a well-defined, reproducible molecular weight and by electrophoretic measurements by Arne Tiselius that indicated that proteins were single molecules.

A second hypothesis, the cyclol hypothesis advanced by Dorothy Wrinch, proposed that the linear polypeptide underwent a chemical cyclol rearrangement C=O + HN C(OH)-N that crosslinked its backbone amide groups, forming a two-dimensional *fabric*. Other primary structures of proteins were proposed by various researchers, such as the diketopiperazine model of Emil Abderhalden and the pyrrol/piperidine model of Troensegaard in 1942. Although never given much credence, these alternative models were finally disproved when Frederick Sanger successfully sequenced insulin and by the crystallographic determination of myoglobin and hemoglobin by Max Perutz and John Kendrew.

Nucleic Acid Sequence

The sequence or primary structure of a nucleic acid is the exact specification of its atomic composition and the chemical bonds connecting those atoms. As nucleic acids, e.g. DNA and RNA, are unbranched polymers, this is equivalent to specifying exact sequence of nucleotides that comprise the whole molecule. This sequence is written as a succession of letters representing a real or hypothetical DNA molecule or strand. By convention, the primary structure of a DNA or RNA molecule is reported from the 5' end to the 3' end.

The sequence has capacity to carry information. When used in reference to biological DNA, which carries the information which directs the functions of living beings, the term genetic sequence is often used. Sequences can be read from the biological raw material through DNA sequencing methods.

Primary structure is sometimes mistakenly termed *primary sequence*, but there is no such term, as well as no parallel concept of secondary or tertiary sequence.

Nucleotides

Nucleic acids consist of a chain of linked units called nucleotides. Each nucleotide consists of three subunits: a phosphate group and a sugar (ribose in the case of RNA, deoxyribose in DNA) make up the backbone of the nucleic acid strand, and attached to the sugar is one of a set of nucleobases. The nucleobases are important in base pairing of strands to form higher-level secondary and tertiary structure such as the famed double helix.

The possible letters are *A*, *C*, *G*, and *T*, representing the four nucleotide bases of a DNA strand — adenine, cytosine, guanine, thymine — covalently linked to a phosphodiester backbone. In the typical case, the sequences are printed abutting one another without gaps, as in the sequence AAAGTCTGAC, read left to right in the 5' to 3' direction. With regards to transcription, a sequence is on the coding strand if it has the same order as the transcribed RNA.

One sequence can be complementary to another sequence, meaning that they have the base on each position is the complementary (i.e. A to T, C to G) and in the reverse order. For example, the complementary sequence to TTAC is GTAA. If one strand of the double-stranded DNA is considered the sense strand, then the other strand, considered the antisense strand, will have the complementary sequence to the sense strand.

Notation

While A, T, C, and G represent a particular nucleotide at a position, there are also letters that represent ambiguity. Of all the molecules sampled, there is more than one kind of nucleotide at that position. The rules of the International Union of Pure and Applied Chemistry (IUPAC) are as follows:

- A = adenine
- C = cytosine
- G = guanine
- T = thymine
- R = G A (purine)
- Y = T C (pyrimidine)
- K = G T (keto)
- M = A C (amino)
- S = G C (strong bonds)
- W = A T (weak bonds)
- B = G T C (all but A)
- D = G A T (all but C)
- H = A C T (all but G)
- V = G C A (all but T)
- N = A G C T (any).

These symbols are also valid for RNA, except with U (uracil) replacing T (thymine).

Apart from adenine (A), cytosine (C), guanine (G), thymine (T) and uracil (U), DNA and RNA also contain bases that have been modified after the nucleic acid chain has been formed. In DNA, the most common modified base is 5-methylcytidine (m5C). In RNA, there are many modified bases, including pseudouridine (ψ), dihydrouridine (D), inosine (I), ribothymidine (rT) and 7-methylguanosine (m7G). Hypoxanthine and xanthine are two of the many bases created through mutagen presence, both of them through deamination (replacement of the amine-group with a carbonyl-group). Hypoxanthine is produced from adenine, xanthine from guanine. Similarly, deamination of cytosine results in uracil.

Biological Significance

In biological systems, nucleic acids contain information which is used by a living cell to construct specific proteins. The sequence of nucleobases on a nucleic acid strand is translated by cell machinery into a sequence of amino acids making up a protein strand. Each group of three bases, called a codon, corresponds to a single amino acid, and there is a specific genetic code by which each possible combination of three bases corresponds to a specific amino acid.

The central dogma of molecular biology outlines the mechanism by which proteins are constructed using information contained in nucleic acids. DNA is transcribed into mRNA molecules, which travels to the ribosome where the mRNA is used as a template for the construction of the protein strand. Since nucleic acids can bind to molecules with complementary sequences, there is a distinction between "sense" sequences which code for proteins, and the complementary "antisense" sequence which is by itself nonfunctional, but can bind to the sense strand.

Sequence Determination

DNA sequencing is the process of determining the nucleotide sequence of a given DNA fragment. The sequence of DNA encodes the necessary information for living things to survive and reproduce. Determining the sequence is therefore useful in fundamental research into why and how organisms live, as well as in applied subjects. Because of the key nature of DNA to living things, knowledge of DNA sequence may come in useful in practically any biological research. For example, in medicine it can be used to identify, diagnose and potentially develop treatments for genetic diseases. Similarly, research into pathogens may lead to treatments for contagious diseases. Biotechnology is a burgeoning discipline, with the potential for many useful products and services. RNA is not sequenced directly. Instead, it is copied to a DNA by reverse transcriptase, and this DNA is then sequenced. Current sequencing methods rely on the discriminatory ability of DNA polymerases, and can therefore only distinguish four bases. An inosine (created from adenosine during RNA editing) will be read as a G, and 5-methyl-cytosine (created from cytosine by DNA methylation) will be read as a C. It is also currently difficult to sequence small amounts of DNA, as the signal will be too weak to measure. This is overcome by PCR amplification.

Digital Format

Once a nucleic acid sequence has been obtained from an organism, it is stored *in silico* in digital format. Digital genetic sequences may be stored in sequence databases, be analysed, be digitally altered and/

or be used as templates for creating new actual DNA using artificial gene synthesis.

Sequence Analysis

The term "sequence analysis" in biology implies subjecting a DNA or peptide sequence to sequence alignment, sequence databases, repeated sequence searches, or other bioinformatics methods on a computer.

Since the development of methods of high-throughput production of gene and protein sequences during the 90s, the rate of addition of new sequences to the databases increases continuously. Such a collection of sequences does not, by itself, increase the scientist's understanding of the biology of organisms. However, comparing sequences with known functions with these new sequences is one way of understanding the biology of that organism from which the new sequence comes.

Thus, sequence analysis can be used to assign function to genes and proteins by the study of the similarities between the compared sequences. Nowadays there are many tools and techniques that provide the sequence comparisons (sequence alignment) and analyse the alignment product to understand the biology.

Sequence analysis in molecular biology and bioinformatics is an automated, computer-based examination of characteristic fragments, e.g. of a DNA strand. It basically includes relevant topics:

1. The comparison of sequences in order to find similarity and dissimilarity in compared sequences (sequence alignment)

2. Identification of gene-structures, reading frames, distributions of introns and exons and regulatory elements

3. Finding and comparing point mutations or the single nucleotide polymorphism (SNP) in organism in order to get the genetic marker.

4. Revealing the evolution and genetic diversity of organisms.

5. Function annotation of genes.

In chemistry, sequence analysis comprises techniques used to do determine the sequence of a polymer formed of several monomers. In molecular biology and genetics, the same process is called simply "sequencing". In marketing, sequence analysis is often used in analytical customer relationship management applications, such as NPTB models (Next Product to Buy).

Methodology

For sequence alignment method compose of pairwise alignment (align with two sequences) and multiple alignment (align with more than two sequence). There are several tools for alignment, including: ClustalW, PROBCONS, MUSCLE, MAFFT, DIALIGN, T-Coffee, POA, and MANGO.

There are many algorithms used in the study of sequence analysis. These include: dynamic programming, Hidden Markov Model, Viterbi and Greedy.

Genetic Testing

The DNA in an organism's genome can be analysed to diagnose vulnerabilities to inherited diseases, and can also be used to determine a child's paternity (genetic father) or a person's ancestry. Normally, every person carries two copies of every gene, one inherited from their mother, one inherited from their father. The human genome is believed to contain around 20,000 - 25,000 genes. In addition to studying chromosomes to the level of individual genes, genetic testing in a broader sense includes biochemical tests for the possible presence of genetic diseases, or mutant forms of genes associated with increased risk of developing genetic disorders.

Genetic testing identifies changes in chromosomes, genes, or proteins. Most of the time, testing is used to find changes that are associated with inherited disorders. The results of a genetic test can confirm or rule out a suspected genetic condition or help determine a person's chance of developing or passing on a genetic disorder. Several hundred genetic tests are currently in use, and more are being developed.

Sequence Alignment

In bioinformatics, a sequence alignment is a way of arranging the sequences of DNA, RNA, or protein to identify regions of similarity that may be a consequence of functional, structural, or evolutionary relationships between the sequences. Aligned sequences of nucleotide or amino acid residues are typically represented as rows within a matrix. Gaps are inserted between the residues so that identical or similar characters are aligned in successive columns. A sequence alignment, produced by ClustalW, of two human zinc finger proteins, identified on the left by GenBank accession number.

Key: Single letters: amino acids. Red: small, hydrophobic, aromatic, not Y. Blue: acidic. Magenta: basic. Green: hydroxyl, amine, amide, basic. Gray: others. "*": identical. ":": conserved substitutions (same colour group). ".": semi-conserved substitution (similar shapes).

Sequence alignments are also used for non-biological sequences, such as those present in natural language or in financial data.

Interpretation

If two sequences in an alignment share a common ancestor, mismatches can be interpreted as point mutations and gaps as indels (that is, insertion or deletion mutations) introduced in one or both lineages in the time since they diverged from one another. In sequence alignments of proteins, the degree of similarity between amino acids occupying a particular position in the sequence can be interpreted as a rough measure of how conserved a particular region or sequence motif is among lineages. The absence of substitutions, or the presence of only very conservative substitutions (that is, the substitution of amino acids whose side chains have similar biochemical properties) in a particular region of the sequence, suggest that this region has structural or functional importance. Although DNA and RNA nucleotide bases are more similar to each other than are amino acids, the conservation of base pairs can indicate a similar functional or structural role.

Alignment Methods

Very short or very similar sequences can be aligned by hand. However, most interesting problems require the alignment of lengthy, highly variable or extremely numerous sequences that cannot be aligned solely by human effort. Instead, human knowledge is applied in constructing algorithms to produce high-quality sequence alignments, and occasionally in adjusting the final results to reflect patterns that are difficult to represent algorithmically (especially in the case of nucleotide sequences).

Computational approaches to sequence alignment generally fall into two categories: *global alignments* and *local alignments*. Calculating a global alignment is a form of global optimization that "forces" the alignment to span the entire length of all query sequences. By contrast, local alignments identify regions of similarity within long sequences that are often widely divergent overall. Local alignments are often preferable, but can be more difficult to calculate because of the additional challenge of identifying the regions of similarity. A variety of computational algorithms have been applied to the sequence alignment problem, including slow but formally optimizing methods like dynamic programming, and efficient, but not as thorough heuristic algorithms or probabilistic methods designed for large-scale database search.

Representations

Alignments are commonly represented both graphically and in text format. In almost all sequence alignment representations, sequences are written in rows arranged so that aligned residues appear in successive columns. In text formats, aligned columns containing identical or similar characters are indicated with a system of conservation symbols. As in the image above, an asterisk or pipe symbol is used to show identity between two columns; other less common symbols include a colon for conservative substitutions and a period for semiconservative substitutions. Many sequence visualization programs also use colour to display information about the properties of the individual sequence elements; in DNA and RNA sequences, this equates to assigning each nucleotide its own colour.

In protein alignments, such as the one in the image above, colour is often used to indicate amino acid properties to aid in judging the conservation of a given amino acid substitution. For multiple sequences the last row in each column is often the consensus sequence determined by the alignment; the consensus sequence is also often represented in graphical format with a sequence logo in which the size of each nucleotide or amino acid letter corresponds to its degree of conservation. Sequence alignments can be stored in a wide variety of text-based file formats, many of which were originally developed in conjunction with a specific alignment program or implementation. Most web-based tools allow a limited number of input and output formats, such as FASTA format and GenBank format and the output is not easily editable. Several conversion programs are available, READSEQ or EMBOSS having a graphical interfaces or command line interfaces, while several programming packages like BioPerl, BioRuby provide functions to do this.

Global and Local Alignments

Illustration of global and local alignments demonstrating the 'gappy' quality of global alignments that can occur if sequences are insufficiently similar Global alignments, which attempt to align every residue in every sequence, are most useful when the sequences in the query set are similar and of roughly equal size. (This does not mean global alignments cannot end in gaps.) A general global alignment technique is the Needleman-Wunsch algorithm, which is based on dynamic programming. Local alignments are more useful for dissimilar sequences that are suspected to contain regions of similarity or similar sequence motifs within their larger sequence context. The Smith-Waterman algorithm is a general local alignment method also based on dynamic programming. With sufficiently similar sequences, there is no difference between local and global alignments.

Hybrid methods, known as semiglobal or "glocal" methods, attempt to find the best possible alignment that includes the start and end of one or the other sequence. This can be especially useful when the downstream part of one sequence overlaps with the upstream part of the other sequence. In this case, neither global nor local alignment is entirely appropriate: a global alignment would attempt to force the alignment to extend beyond the region of overlap, while a local alignment might not fully cover the region of overlap.

Pairwise Alignment

Pairwise sequence alignment methods are used to find the best-matching piecewise (local) or global alignments of two query sequences. Pairwise alignments can only be used between two sequences at a time, but they are efficient to calculate and are often used for methods that do not require extreme precision (such as searching a database for sequences with high similarity to a query). The three primary methods of producing pairwise alignments are dot-matrix methods, dynamic programming, and word methods; however, multiple sequence alignment techniques can also align pairs of sequences. Although each method has its individual strengths and weaknesses, all three pairwise methods have difficulty with highly repetitive sequences of low information content-especially where the number of repetitions differ in the two sequences to be aligned. One way of quantifying the utility of a given pairwise alignment is the 'maximum unique match', or the longest subsequence that occurs in both query sequence. Longer MUM sequences typically reflect closer relatedness.

Dot-matrix Methods

A DNA dot plot of a human zinc finger transcription factor (GenBank ID NM_002383), showing regional self-similarity. The main diagonal represents the sequence's alignment with itself; lines off the main diagonal represent similar or repetitive patterns within the sequence. This is a typical example of a recurrence plot.

The dot-matrix approach, which implicitly produces a family of alignments for individual sequence regions, is qualitative and conceptually simple, though time-consuming to analyse on a large scale. In the absence of noise, it can be easy to visually identify certain sequence features—such as insertions, deletions, repeats, or inverted repeats—from a dot-matrix plot. To construct a dot-matrix plot, the two sequences are written along the top row and leftmost column of a two-dimensional matrix and a dot is placed at any point where the characters in the appropriate columns match—this is a typical recurrence plot. Some implementations vary the size or intensity of the dot depending on the degree of similarity of the two characters, to accommodate conservative substitutions. The dot plots of very closely related sequences will appear as a single line along the matrix's main diagonal.

Problems with dot plots as an information display technique include: noise, lack of clarity, non-intuitiveness, difficulty extracting match summary statistics and match positions on the two sequences. There is also much wasted space where the match data is inherently duplicated across the diagonal and most of the actual area of the plot is taken up by either empty space or noise, and, finally, dot-plots are limited to two sequences. None of these limitations apply to Miropeats alignment diagrams but they have their own particular flaws.

Dot plots can also be used to assess repetitiveness in a single sequence. A sequence can be plotted against itself and regions that share significant similarities will appear as lines off the main diagonal. This effect can occur when a protein consists of multiple similar structural domains.

Dynamic Programming

The technique of dynamic programming can be applied to produce global alignments via the Needleman-Wunsch algorithm, and local alignments via the Smith-Waterman algorithm. In typical usage, protein alignments use a substitution matrix to assign scores to amino-acid matches or mismatches, and a gap penalty for matching an amino acid in one sequence to a gap in the other. DNA and RNA alignments may use a scoring matrix, but in practice often simply assign a positive match score, a negative mismatch score, and a negative gap penalty. (In standard dynamic programming, the score of each amino acid position is independent of the identity of its neighbours, and therefore base stacking effects are not taken into account. However, it is possible to account for such effects by modifying the algorithm.) A common extension to standard linear gap costs, is the usage of two different gap penalties for opening a gap and for extending a gap. Typically the former is much larger than the latter, e.g. -10 for gap open and -2 for gap extension. Thus, the number of gaps in an alignment is usually reduced and residues and gaps are kept together, which typically makes more biological sense. The Gotoh algorithm implements affine gap costs by using three matrices.

Dynamic programming can be useful in aligning nucleotide to protein sequences, a task complicated by the need to take into account frameshift mutations (usually insertions or deletions). The framesearch method produces a series of global or local pairwise alignments between a query nucleotide sequence and a search set of protein sequences, or vice versa. Its ability to evaluate frameshifts offset by an arbitrary number of nucleotides makes the method useful for sequences containing large numbers of indels, which can be very difficult to align with more efficient heuristic methods. In practice, the method requires large amounts of computing power or a system whose architecture is specialized for dynamic programming. The BLAST and EMBOSS suites provide basic tools for creating translated alignments (though some of these approaches take advantage of side-effects of sequence searching capabilities of the tools). More general methods are available from both commercial sources, such as *FrameSearch*, distributed as part of the Accelrys GCG package, and Open Source software such as Genewise.

The dynamic programming method is guaranteed to find an optimal alignment given a particular scoring function; however, identifying a good scoring function is often an empirical rather than a theoretical matter. Although dynamic programming is extensible to more than two sequences, it is prohibitively slow for large numbers of or extremely long sequences.

Word Methods

Word methods, also known as *k*-tuple methods, are heuristic methods that are not guaranteed to find an optimal alignment solution, but are significantly more efficient than dynamic programming. These methods are especially useful in large-scale database searches where it is understood that a large proportion of the candidate sequences will have essentially no significant match with the query sequence. Word methods are best known for their implementation in the database search tools FASTA and the BLAST family. Word methods identify a series of short, nonoverlapping subsequences ("words") in the query sequence that are then matched to candidate database sequences. The relative positions of the word in the two sequences being compared are subtracted to obtain an offset; this will indicate a region of alignment if multiple distinct words produce the same offset. Only if this region is detected do these methods apply more sensitive alignment criteria; thus, many unnecessary comparisons with sequences of no appreciable similarity are eliminated.

In the FASTA method, the user defines a value k to use as the word length with which to search the database. The method is slower but more sensitive at lower values of k, which are also preferred for searches involving a very short query sequence. The BLAST family of search methods provides a number of algorithms optimized for particular types of queries, such as searching for distantly related sequence matches. BLAST was developed to provide a faster alternative to FASTA without sacrificing much accuracy; like FASTA, BLAST uses a word search of length k, but evaluates only the most significant word matches, rather than every word match as does FASTA. Most BLAST implementations use a fixed default word length that is optimized for the query and database type, and that is changed only under special circumstances, such as when searching with repetitive or very short query sequences. Implementations can be found via a number of web portals, such as EMBL FASTA and NCBI BLAST.

Multiple Sequence Alignment

Multiple sequence alignment is an extension of pairwise alignment to incorporate more than two sequences at a time. Multiple alignment methods try to align all of the sequences in a given query set. Multiple alignments are often used in identifying conserved sequence regions across a group of sequences hypothesized to be evolutionarily related. Such conserved sequence motifs can be used in conjunction with structural and mechanistic information to locate the catalytic active sites of enzymes. Alignments are also used to aid in establishing evolutionary relationships by constructing phylogenetic trees. Multiple sequence alignments are computationally difficult to produce and most formulations of the problem lead to NP-complete combinatorial optimization problems. Nevertheless, the utility of these alignments in bioinformatics has led to the development of a variety of methods suitable for aligning three or more sequences.

Dynamic Programming

The technique of dynamic programming is theoretically applicable to any number of sequences; however, because it is computationally expensive in both time and memory, it is rarely used for more than three or four sequences in its most basic form.

This method requires constructing the n-dimensional equivalent of the sequence matrix formed from two sequences, where n is the number of sequences in the query. Standard dynamic programming is first used on all pairs of query sequences and then the "alignment space" is filled in by considering possible matches or gaps at intermediate positions, eventually constructing an alignment essentially between each two-sequence alignment. Although this technique is computationally expensive, its guarantee of a global optimum solution is useful in cases where only a few sequences need to be aligned accurately. One method for reducing the computational demands of dynamic programming, which relies on the "sum of pairs" objective function, has been implemented in the MSA software package.

Progressive Methods

Progressive, hierarchical, or tree methods generate a multiple sequence alignment by first aligning the most similar sequences and then adding successively less related sequences or groups to the alignment until the entire query set has been incorporated into the solution. The initial tree describing the sequence relatedness is based on pairwise comparisons that may include heuristic pairwise alignment methods similar to FASTA. Progressive alignment results are dependent on the choice of "most related" sequences and thus can be sensitive to inaccuracies in the initial pairwise alignments. Most progressive multiple sequence alignment methods additionally weight the sequences in the query set according to their relatedness, which reduces the likelihood of making a poor choice of initial sequences and thus improves alignment accuracy.

Many variations of the Clustal progressive implementation are used for multiple sequence alignment, phylogenetic tree construction, and as input for protein structure prediction. A slower but more accurate variant of the progressive method is known as T-Coffee.

Iterative Methods

Iterative methods attempt to improve on the weak point of the progressive methods, the heavy dependence on the accuracy of the initial pairwise alignments. Iterative methods optimize an objective function based on a selected alignment scoring method by assigning an initial global alignment and then realigning sequence subsets. The realigned subsets are then themselves aligned to produce the next iteration's multiple sequence alignment. Various ways of selecting the sequence subgroups and objective function are reviewed in.

Motif Finding

Motif finding, also known as profile analysis, constructs global multiple sequence alignments that attempt to align short conserved sequence motifs among the sequences in the query set. This is usually done by first constructing a general global multiple sequence alignment, after which the highly conserved regions are isolated and used to construct a set of profile matrices. The profile matrix for each conserved region is arranged like a scoring matrix but its frequency counts for each amino acid or nucleotide at each position are derived from the conserved region's character distribution rather than from a more general empirical distribution. The profile matrices are then used to search other sequences for occurrences of the motif they characterize. In cases where the original data set contained a small number of sequences, or only highly related sequences, pseudocounts are added to normalize the character distributions represented in the motif.

Techniques Inspired by Computer Science

A variety of general optimization algorithms commonly used in computer science have also been applied to the multiple sequence alignment problem. Hidden Markov models have been used to produce probability scores for a family of possible multiple sequence alignments for a given query set; although early HMM-based methods produced underwhelming performance, later applications have found them especially effective in detecting remotely related sequences because they are less susceptible to noise created by conservative or semiconservative substitutions. Genetic algorithms and simulated annealing have also been used in optimizing multiple sequence alignment scores as judged by a scoring function like the sum-of-pairs method. More complete details and software packages can be found in the main article multiple sequence alignment.

Structural Alignment

Structural alignments, which are usually specific to protein and sometimes RNA sequences, use information about the secondary and tertiary structure of the protein or RNA molecule to aid in aligning the sequences. These methods can be used for two or more sequences and typically produce local alignments; however, because they depend on the availability of structural information, they can only be used for sequences whose corresponding structures are known (usually through X-ray crystallography or NMR spectroscopy). Because both protein and RNA structure is more evolutionarily conserved than sequence, structural alignments can be more reliable between sequences that are very distantly related and that have diverged so extensively that sequence comparison cannot reliably detect their similarity.

Structural alignments are used as the "gold standard" in evaluating alignments for homology-based protein structure prediction because they explicitly align regions of the protein sequence that are structurally similar rather than relying exclusively on sequence information. However, clearly structural alignments cannot be used in structure prediction because at least one sequence in the query set is the target to be modelled, for which the structure is not known. It has been shown that, given the structural alignment between a target and a template sequence, highly accurate models of the target protein sequence can be produced; a major stumbling block in homology-based structure prediction is the production of structurally accurate alignments given only sequence information.

DALI

The DALI method, or distance matrix alignment, is a fragment-based method for constructing structural alignments based on contact similarity patterns between successive hexapeptides in the query sequences. It can generate pairwise or multiple alignments and identify a query sequence's structural neighbours in the Protein Data Bank (PDB). It has been used to construct the FSSP structural alignment database (Fold classification based on Structure-Structure alignment of Proteins, or Families of Structurally Similar Proteins). A DALI webserver can be accessed at EBI DALI and the FSSP is located at The Dali Database.

SSAP

SSAP (sequential structure alignment program) is a dynamic programming-based method of structural alignment that uses atom-to-atom vectors in structure space as comparison points. It has been extended since its original description to include multiple as well as pairwise alignments, and has been used in the construction of the CATH (Class, Architecture, Topology, Homology) hierarchical database classification of protein folds. The CATH database can be accessed at CATH Protein Structure Classification.

Combinatorial Extension

The combinatorial extension method of structural alignment generates a pairwise structural alignment by using local geometry to align short fragments of the two proteins being analysed and then assembles these fragments into a larger alignment. Based on measures such as rigid-body root mean square distance, residue distances, local secondary structure, and surrounding environmental features such as residue neighbour hydrophobicity, local alignments called "aligned fragment pairs" are generated and used to build a similarity matrix representing all possible structural alignments within predefined cutoff criteria. A path from one protein structure state to the other is then traced through the matrix by extending the growing alignment one fragment at a time. The optimal such path defines the combinatorial-extension alignment. A web-based server implementing the method and providing a database of pairwise alignments of structures in the Protein Data Bank is located at the Combinatorial Extension website.

Phylogenetic Analysis

Phylogenetics and sequence alignment are closely related fields due to the shared necessity of evaluating sequence relatedness. The field of phylogenetics makes extensive use of sequence alignments in the construction and interpretation of phylogenetic trees, which are used to classify the evolutionary relationships between homologous genes represented in the genomes of divergent species. The degree to which sequences in a query set differ is qualitatively related to the sequences' evolutionary distance from one another. Roughly speaking, high sequence identity suggests that the sequences in question have a comparatively young most recent common ancestor, while low identity suggests that the divergence is more ancient. This approximation, which reflects the "molecular clock" hypothesis that a roughly constant rate of evolutionary change can be used to extrapolate the elapsed time since two genes first diverged (that is, the coalescence time), assumes that the effects of mutation and selection are constant across sequence lineages. Therefore it does not account for possible difference among organisms or species in the rates of DNA repair or the possible functional conservation of specific regions in a sequence. (In the case of nucleotide sequences, the molecular clock hypothesis in its most basic form also discounts the difference in acceptance rates between silent mutations that do not alter the meaning of a given codon and other mutations that result in a different amino acid being incorporated into the protein.) More statistically accurate methods allow the evolutionary rate on each branch of the phylogenetic tree to vary, thus producing better estimates of coalescence times for genes.

Progressive multiple alignment techniques produce a phylogenetic tree by necessity because they incorporate sequences into the growing alignment in order of relatedness. Other techniques that assemble multiple sequence alignments and phylogenetic trees score and sort trees first and calculate a multiple sequence alignment from the highest-scoring tree. Commonly used methods of phylogenetic tree construction are mainly heuristic because the problem of selecting the optimal tree, like the problem of selecting the optimal multiple sequence alignment, is NP-hard.

Assessment of Significance

Sequence alignments are useful in bioinformatics for identifying sequence similarity, producing phylogenetic trees, and developing homology models of protein structures. However, the biological relevance of sequence alignments is not always clear. Alignments are often assumed to reflect a degree of evolutionary change between sequences descended from a common ancestor; however, it is formally possible that convergent evolution can occur to produce apparent similarity between proteins that are evolutionarily unrelated but perform similar functions and have similar structures.

In database searches such as BLAST, statistical methods can determine the likelihood of a particular alignment between sequences or sequence regions arising by chance given the size and composition of the database being searched. These values can vary significantly depending on the search space. In particular, the likelihood of finding a given alignment by chance increases if the database consists only of sequences from the same organism as the query sequence. Repetitive sequences in the database or query can also distort both the search results and the assessment of statistical significance; BLAST automatically filters such repetitive sequences in the query to avoid apparent hits that are statistical artifacts.

Methods of statistical significance estimation for gapped sequence alignments are available in the literature.

Assessment of Credibility

Statistical significance indicates the probability that an alignment of a given quality could arise by chance, but does not indicate how much superior a given alignment is to alternative alignments of the same sequences. Measures of alignment credibility indicate the extent to which the best scoring alignments for a given pair of sequences are substantially similar. Methods of alignment credibility estimation for gapped sequence alignments are available in the literature.

Scoring Functions

The choice of a scoring function that reflects biological or statistical observations about known sequences is important to producing good alignments. Protein sequences are frequently aligned using substitution matrices that reflect the probabilities of given character-to-character substitutions. A series of matrices called PAM matrices (Point Accepted

Mutation matrices, originally defined by Margaret Dayhoff and sometimes referred to as "Dayhoff matrices") explicitly encode evolutionary approximations regarding the rates and probabilities of particular amino acid mutations. Another common series of scoring matrices, known as BLOSUM (Blocks Substitution Matrix), encodes empirically derived substitution probabilities. Variants of both types of matrices are used to detect sequences with differing levels of divergence, thus allowing users of BLAST or FASTA to restrict searches to more closely related matches or expand to detect more divergent sequences. Gap penalties account for the introduction of a gap- on the evolutionary model, an insertion or deletion mutation- in both nucleotide and protein sequences, and therefore the penalty values should be proportional to the expected rate of such mutations. The quality of the alignments produced therefore depends on the quality of the scoring function.

It can be very useful and instructive to try the same alignment several times with different choices for scoring matrix and/or gap penalty values and compare the results. Regions where the solution is weak or non-unique can often be identified by observing which regions of the alignment are robust to variations in alignment parameters.

Other Biological Uses

Sequenced RNA, such as expressed sequence tags and full-length mRNAs, can be aligned to a sequenced genome to find where there are genes and get information about alternative splicing and RNA editing. Sequence alignment is also a part of genome assembly, where sequences are aligned to find overlap so that *contigs* (long stretches of sequence) can be formed. Another use is SNP analysis, where sequences from different individuals are aligned to find single basepairs that are often different in a population.

Non-biological Uses

The methods used for biological sequence alignment have also found applications in other fields, most notably in natural language processing and in social sciences. Techniques that generate the set of elements from which words will be selected in natural-language generation algorithms have borrowed multiple sequence alignment techniques from bioinformatics to produce linguistic versions of computer-generated mathematical proofs. In the field of historical and comparative linguistics, sequence alignment has been used to partially automate the comparative method by which linguists traditionally reconstruct languages. Business and marketing research has also applied multiple sequence alignment techniques in analysing series of purchases over time.

Software

A more complete list of available software categorized by algorithm and alignment type is available at sequence alignment software, but common software tools used for general sequence alignment tasks include ClustalW and T-coffee for alignment, and BLAST and FASTA3x for database searching.

Alignment algorithms and software can be directly compared to one another using a standardized set of benchmark reference multiple sequence alignments known as BAliBASE. The data set consists of structural alignments, which can be considered a standard against which purely sequence-based methods are compared. The relative performance of many common alignment methods on frequently encountered alignment problems has been tabulated and selected results published online at BAliBASE. A comprehensive list of BAliBASE scores for many (currently 12) different alignment tools can be computed within the protein workbench STRAP.

Sequence Motif

In genetics, a sequence motif is a nucleotide or amino-acid sequence pattern that is widespread and has, or is conjectured to have, a biological significance. For proteins, a sequence motif is distinguished from a structural motif, a motif formed by the three dimensional arrangement of amino acids, which may not be adjacent.

An example is the *N*-glycosylation site motif:

Asn, followed by anything but Pro, followed by either Ser or Thr, followed by anything but Pro.

Where the three-letter abbreviations are the conventional designations for amino acids.

When a sequence motif appears in the exon of a gene, it may encode the "structural motif" of a protein; that is a stereotypical element of the overall structure of the protein. Nevertheless, motifs need not be associated with a distinctive secondary structure. "Noncoding" sequences are not translated into proteins, and nucleic acids with such motifs need not deviate from the typical shape (e.g. the "B-form" DNA double helix).

Outside of gene exons, there exist regulatory sequence motifs and motifs within the "junk," such as satellite DNA. Some of these are believed to affect the shape of nucleic acids, but this is only sometimes the case. For example, many DNA binding proteins that have affinities for specific motifs only bind DNA in its double-helical form. They are able to recognize motifs through contact with the double helix's major or minor groove.

Short coding motifs, which appear to lack secondary structure, include those that label proteins for delivery to particular parts of a cell, or mark them for phosphorylation. Within a sequence or database of sequences, researchers search and find motifs using computer-based techniques of sequence analysis, such as BLAST. Such techniques belong to the discipline of bioinformatics.

Motif Bioinformatics

Consider the N-glycosylation site motif mentioned above:

Asn, followed by anything but Pro, followed by either Ser or Thr, followed by anything but Pro.

This pattern may be written as N{P}[ST]{P} where N = Asn, P = Pro, S = Ser, T = Thr; {X} means any amino acid except X; and [XY] means either X or Y.

The notation [XY] does not give any indication of the probability of X or Y occurring in the pattern. Sometimes patterns are defined in terms of a probabilistic model such as a hidden Markov model.

Motifs and Consensus Sequences

The notation [XYZ] means X or Y or Z, but does not indicate the likelihood of any particular match. For this reason, two or more patterns are often associated with a single motif: the defining pattern, and various typical patterns.

For example, the defining sequence for the IQ motif may be taken to be:

[FILV]Qxxx[RK]Gxxx[RK]xx[FILVWY]

Where x signifies any amino acid, and the square brackets indicate an alternative.

Usually, however, the first letter is I, and both [RK] choices resolve to R. Since the last choice is so wide, the pattern IQxxxRGxxxR is sometimes equated with the IQ motif itself, but a more accurate description would be a *consensus sequence for the IQ motif*.

De Novo Computational Discovery of Motifs

There are software programs which, given multiple input sequences, attempt to identify one or more candidate motifs. One example is MEME, which generates statistical information for each candidate. Other algorithms include CisModule, AlignAce, PhyloGibbs, Weeder, Amadeus and FIRE. SCOPE is an ensemble motif finder that uses several algorithms simultaneously. There currently exist more than 100 publications with similar algorithms without a comprehensive benchmark so selecting one is not straightforward.

Discovery through Evolutionary Conservation

Motifs have been discovered by studying similar genes in different species. For example, by aligning the amino acid sequences specified by the GCM (*glial cells missing*) gene in man, mouse and *D. melanogaster*, Akiyama and others discovered a pattern which they called the GCM motif. It spans about 150 amino acid residues, and begins as follows:

WDIND*.*P..*...D.F.*W***.**.IYS**...A.*H*S*WAMRNTNNHN

Here each. signifies a single amino acid or a gap, and each * indicates one member of a closely-related family of amino acids.

The authors were able to show that the motif has DNA binding activity. A motif discovery algorithm that considers phylogenetic conservation is PhyloGibbs.

Pattern Description Notations

Several notations for describing motifs are in use but most of them are variants of standard notations for regular expressions and use these conventions:

- there is an alphabet of single characters, each denoting a specific amino acid or a set of amino acids;
- a string of characters drawn from the alphabet denotes a sequence of the corresponding amino acids;
- any string of characters drawn from the alphabet enclosed in square brackets matches any one of the corresponding amino acids; e.g. [abc] matches any of the amino acids represented by a or b or c.

The fundamental idea behind all these notations is the matching principle, which assigns a meaning to a sequence of elements of the pattern notation:

A sequence of elements of the pattern notation matches a sequence of amino acids if and only if the latter sequence can be partitioned into subsequences in such a way that each pattern element matches the corresponding subsequence in turn.

Thus the pattern [AB] [CDE] F matches the six amino acid sequences corresponding to ACF, ADF, AEF, BCF, BDF, and BEF. Different pattern description notations have other ways of forming pattern elements. One of these notations is the PROSITE notation, described in the following subsection.

Prosite Pattern Notation

The Prosite notation uses the IUPAC one-letter codes and conforms to the above description with the exception that a concatenation symbol, '-', is used between pattern elements, but it is often dropped between letters of the pattern alphabet. Prosite allows the following pattern elements in addition to those described previously:

- The lower case letter 'x' can be used as a pattern element to denote any amino acid.

- A string of characters drawn from the alphabet and enclosed in braces (curly brackets) denotes any amino acid except for those in the string. For example, {ST} denotes any amino acid other than S or T.

- If a pattern is restricted to the N-terminal of a sequence, the pattern is prefixed with '<'.

- If a pattern is restricted to the C-terminal of a sequence, the pattern is suffixed with '>'.

- The character '>' can also occur inside a terminating square bracket pattern, so that S[T>] matches both "ST" and "S>".

- If e is a pattern element, and m and n are two decimal integers with m <= n, then:
  — e(m) is equivalent to the repetition of e exactly m times;
  — e(m,n) is equivalent to the repetition of e exactly k times for any integer k satisfying: m <= k <= n.

Some examples:

- x(3) is equivalent to x-x-x.
- x(2,4) matches any sequence that matches x-x or x-x-x or x-x-x-x.

The signature of the C2H2-type *zinc finger* domain is:

- C-x(2,4)-C-x(3)-[LIVMFYWC]-x(8)-H-x(3,5)-H

Matrices

A matrix of numbers containing scores for each residue or nucleotide at each position of a fixed-length motif. There are two types of weight matrices.

- A position frequency matrix (PFM) records the position-dependent frequency of each residue or nucleotide. PFMs can be experimentally determined from SELEX experiments or computationally discovered by tools such as MEME using hidden Markov models.

- A position weight matrix (PWM) contains log odds weights for computing a match score. A cutoff is needed to specify whether an input sequence matches the motif or not. PWMs are calculated from PFMs.

Another Scheme

The following example comes from the paper by Matsuda, et al. 1997.

The *E. coli* lactose operon repressor LacI (PDB 1lcc chain A) and *E. coli* catabolite gene activator (PDB 3gap chain A) both have a *helix-turn-helix* motif, but their amino acid sequences do not show much similarity. Matsuda, et al. devised a code they called the "three-dimensional chain code" for representing a protein structure as a string of letters. This encoding scheme reveals the similarity between the proteins much more clearly than the amino acid sequence:

Biomolecular Structure

Biomolecular structure is the structure of biomolecules, mainly proteins and the nucleic acids DNA and RNA. The structure of these molecules is frequently decomposed into primary structure, secondary structure, tertiary structure, and quaternary structure. The scaffold for this structure is provided by secondary structural elements which are hydrogen bonds within the molecule. This leads to several recognizable "domains" of protein structure and nucleic acid structure, including secondary structure like hairpin loops, bulges and internal loops for nucleic acids, and alpha helices and beta sheets for proteins.

The terms primary, secondary, tertiary, and quaternary structure were first coined by Kaj Ulrik Linderstrom-Lang in his 1951 Lane Medical Lectures at Stanford University.

Primary Structure

In biochemistry, the Primary Structure of a biological molecule is the exact specification of its atomic composition and the chemical bonds connecting those atoms (including stereochemistry). For a typical unbranched, un-crosslinked biopolymer (such as a molecule of DNA, RNA or typical intracellular protein), the primary structure is equivalent to specifying the sequence of its monomeric subunits, e.g., the nucleotide or peptide sequence.

Primary structure is sometimes mistakenly termed *primary sequence*, but there is no such term, as well as no parallel concept of secondary or tertiary sequence. By convention, the primary structure of a protein is reported starting from the amino-terminal (N) end to the carboxyl-terminal (C) end, while the primary structure of DNA or RNA molecule is reported from the 5' end to the 3' end.

The primary structure of a nucleic acid molecule refers to the exact sequence of nucleotides that comprise the whole molecule. Frequently the primary structure encodes motifs that are of functional importance. Some examples of sequence motifs are: the C/D and H/ ACA boxes of snoRNAs, Sm binding site found in spliceosomal RNAs such as U1, U2, U4, U5, U6, U12 and U3, the Shine-Dalgarno sequence, the Kozak consensus sequence and the RNA polymerase III terminator.

Secondary Structure

Secondary (inset) and tertiary structure of tRNA demonstrating coaxial stacking. PDB (6tna) rendered via PyMOL.

In biochemistry and structural biology, secondary structure is the general three-dimensional form of *local segments* of biopolymers such as proteins and nucleic acids (DNA/RNA). It does not, however, describe specific atomic positions in three-dimensional space, which are considered to be tertiary structure. Secondary structure is formally defined by the hydrogen bonds of the biopolymer, as observed in an atomic-resolution structure. In proteins, the secondary structure is defined by patterns of hydrogen bonds between backbone amide and carboxyl groups (sidechain-mainchain and sidechain-sidechain hydrogen bonds are irrelevant), where the DSSP definition of a hydrogen bond is used. In nucleic acids, the secondary structure is defined by the hydrogen bonding between the nitrogenous bases.

For proteins, however, the hydrogen bonding is correlated with other structural features, which has given rise to less formal definitions of secondary structure. For example, residues in protein helices generally adopt backbone dihedral angles in a particular region of the Ramachandran plot; thus, a segment of residues with such dihedral angles is often called a "helix", regardless of whether it has the correct hydrogen bonds. Many other less formal definitions have been proposed, often applying concepts from the differential geometry of curves, such as curvature and torsion. Least formally, structural biologists solving a new atomic-resolution structure will sometimes assign its secondary structure "by eye" and record their assignments in the corresponding PDB file.

The secondary structure of a nucleic acid molecule refers to the basepairing interactions within a single molecule or set of interacting molecules. The secondary structure of biological RNA's can often be uniquely decomposed into stems and loops. Frequently these elements, or combinations of them, can be further classified, for example, tetraloops, pseudoknots and stem-loops. There are many secondary structure elements of functional importance to biological RNA's; some famous examples are the Rho-independent terminator stem-loops and the tRNA cloverleaf. There is a minor industry of researchers attempting to determine the secondary structure of RNA molecules. Approaches include both experimental and computational methods.

Tertiary Structure

In biochemistry and molecular biology, the tertiary structure of a protein or any other macromolecule is its three-dimensional structure, as defined by the atomic coordinates. Proteins and nucleic acids are capable of diverse functions ranging from molecular recognition to catalysis. Such functions require a precise three-dimensional tertiary structure. While such structures are diverse and seemingly complex, they are composed of recurring, easily recognizable tertiary structure motifs that serve as molecular building blocks. Tertiary structure is considered to be largely determined by the biomolecule's primary structure, or the sequence of amino acids or nucleotides of which it is composed. Efforts to predict tertiary structure from the primary structure are known generally as structure prediction.

Quaternary Structure

In biochemistry, quaternary structure is the arrangement of multiple folded protein or coiling protein molecules in a multi-subunit complex. For nucleic acids, the term is less common, but can refer to the higher-level organization of DNA in chromatin, including its interactions with histones, or to the interactions between separate RNA units in the ribosome or spliceosome.

Structure Determination

Structure probing is the process by which biochemical techniques are used to determine biomolecular structure. This analysis can be used to define the patterns which can infer the molecular structure, experimental analysis of molecular structure and function, and further understanding on development of smaller molecules for further biological research. Structure probing analysis can be done through many different methods, which include chemical probing, hydroxyl radical probing, nucleotide analog interference mapping (NAIM), and in-line probing.

DNA structures can be determined using either nuclear magnetic resonance spectroscopy or X-ray crystallography. The first published reports of A-DNA X-ray diffraction patterns— and also B-DNA— employed analyses based on Patterson transforms that provided only a limited amount of structural information for oriented fibres of DNA isolated from calf thymus. An alternate analysis was then proposed by Wilkins et al. in 1953 for B-DNA X-ray diffraction/scattering patterns of hydrated, bacterial oriented DNA fibres and trout sperm heads in terms of squares of Bessel functions. Although the 'B-DNA form' is most common under the conditions found in cells, it is not a well-defined conformation but a family or fuzzy set of DNA-conformations that occur at the high hydration levels present in a wide variety of living cells. Their corresponding X-ray diffraction & scattering patterns are characteristic of molecular paracrystals with a significant degree of disorder (>20%), and concomitantly the structure is not tractable using only the standard analysis.

On the other hand, the standard analysis, involving only Fourier transforms of Bessel functions and DNA molecular models, is still routinely employed for the analysis of A-DNA and Z-DNA X-ray diffraction patterns.

Structure Prediction

S. cerevisiae tRNA-PHE structure space: the energies and structures were calculated using RNAsubopt and the structure distances computed using RNAdistance.

Main articles: Protein structure prediction and Nucleic acid structure prediction.

Biomolecular structure prediction is the prediction of the three-dimensional structure of a protein from its amino acid sequence, or of a nucleic acid from its base sequence. In other words, it is the prediction of secondary and tertiary structure from its primary structure. Structure prediction is the inverse of biomolecular design.

Protein structure prediction is one of the most important goals pursued by bioinformatics and theoretical chemistry. Protein structure prediction is of high importance in medicine (for example, in drug design) and biotechnology (for example, in the design of novel enzymes).

Every two years, the performance of current methods is assessed in the CASP experiment. There has also been a significant amount of bioinformatics research directed at the RNA structure prediction problem. A common problem for researchers working with RNA is to determine the three-dimensional structure of the molecule given just the nucleic acid sequence. However, in the case of RNA much of the final structure is determined by the secondary structure or intra-molecular base-pairing interactions of the molecule. This is shown by the high conservation of base-pairings across diverse species.

Secondary structure of small nucleic acid molecules is largely determined by strong, local interactions such as hydrogen bonds and base stacking. Summing the free energy for such interactions, usually using a nearest-neighbour model, provides an approximation for the stability of given structure. The most straighforward way to find the lowest free energy structure would be to generate all possible structures and calculate the free energy for it, but the number of possible structures for a sequence increases exponentially with the length of the nucleic acid. For longer molecules, the number of possible secondary structures is enormous.

Sequence covariation methods rely on the existence of a data set composed of multiple homologous RNA sequences with related but dissimilar sequences. These methods analyse the covariation of individual base sites in evolution; maintenance at two widely separated sites of a pair of base-pairing nucleotides indicates the presence of a structurally required hydrogen bond between those positions. The general problem of pseudoknot prediction has been shown to be NP-complete.

Design

Biomolecular design can be considered the inverse of structure prediction. In structure prediction, the structure is determined from a known sequence, while in nucleic acid design, a sequence is generated which will form a desired structure.

Other Biomolecules

Other biomolecules, such as polysaccharides and lipids, can also have higher-order structure of biological consequence.

Non-standard Amino Acids

Aside from the twenty-two standard amino acids, there are a vast number of "non-standard" amino acids. These non-standard amino acids found in proteins are formed by post-translational modification, which is modification after translation in protein synthesis. These modifications are often essential for the function or regulation of a protein; for example, the carboxylation of glutamate allows for better binding of calcium cations, and the hydroxylation of proline is critical for maintaining connective tissues. Another example is the formation of hypusine in the translation initiation factor EIF5A, through modification of a lysine residue. Such modifications can also determine the localization of the protein, e.g., the addition of long hydrophobic groups can cause a protein to bind to a phospholipid membrane.

Examples of nonstandard amino acids that are not found in proteins include lanthionine, 2-aminoisobutyric acid, dehydroalanine and the neurotransmitter gamma-aminobutyric acid. Nonstandard amino acids often occur as intermediates in the metabolic pathways for standard amino acids — for example ornithine and citrulline occur in the urea cycle, part of amino acid catabolism. A rare exception to the dominance of α-amino acids in biology is the β-amino acid beta alanine (3-aminopropanoic acid), which is used in plants and microorganisms in the synthesis of pantothenic acid (vitamin B_5), a component of coenzyme A.

Amino Acid Synthesis

Amino acid synthesis is the set of biochemical processes (metabolic pathways) by which the various amino acids are produced from other compounds. The substrates for these processes are various compounds in the organism's diet or growth media. Not all organisms are able to synthesise all amino acids, for example humans are only able to synthesise 12 of the 20 standard amino acids.

A fundamental problem for biological systems is to obtain nitrogen in an easily usable form. This problem is solved by certain microorganisms capable of reducing the inert Na"N molecule (nitrogen gas) to two molecules of ammonia in one of the most remarkable reactions in biochemistry. Nitrogen in the form of ammonia is the source of nitrogen for all the amino acids. The carbon backbones come from the glycolytic pathway, the pentose phosphate pathway, or the citric acid cycle. In amino acid production, one encounters an important problem in biosynthesis — namely, stereochemical control. Because all amino acids except glycine are chiral, biosynthetic pathways must generate the correct isomer with high fidelity. In each of the 19 pathways for the generation of chiral amino acids, the stereochemistry at the α-carbon atom is established by a transamination reaction that involves pyridoxal phosphate. Almost all the transaminases that catalyse these reactions descend from a common ancestor, illustrating once again that effective solutions to biochemical problems are retained throughout evolution.

Biosynthetic pathways are often highly regulated such that building blocks are synthesized only when supplies are low. Very often, a high concentration of the final product of a pathway inhibits the activity of enzymes that function early in the pathway. Often present are allosteric enzymes capable of sensing and responding to concentrations of regulatory species. These enzymes are similar in functional properties to aspartate transcarbamoylase and its regulators. Feedback and allosteric mechanisms ensure that all twenty amino acids are maintained in sufficient amounts for protein synthesis and other processes.

Amino Acid Synthesis

Amino acids are synthesized from α-ketoacids, and later transaminated from another aminoacid, usually Glutamate. The enzyme involved in this reaction is an aminotransferase.

Nitrogen Fixation: Microorganisms Use ATP and a Powerful Reductant to Reduce Atmospheric Nitrogen to Ammonia

Microorganisms use ATP and reduced ferredoxin, a powerful reductant, to reduce N2 to NH3. An iron-molybdenum cluster in nitrogenase deftly catalyses the fixation of N2, a very inert molecule. Higher organisms consume the fixed nitrogen to synthesize amino acids, nucleotides, and other nitrogen-containing biomolecules. The major points of entry of NH4+ into metabolism are glutamine or glutamate.

Amino Acids are Made from Intermediates of the Citric Acid Cycle and Other Major Pathways

Of the basic set of 20 amino acids (not counting selenocysteine) there are 8 that human beings cannot synthesize. In addition, the amino acids arginine, cysteine, glycine, glutamine, histidine, proline, serine and tyrosine are considered conditionally essential, meaning they are not normally required in the diet, but must be supplied exogenously to specific populations that do not synthesize it in adequate amounts. (For example, enough arginine is synthesized by the urea cycle to meet the needs of an adult but perhaps not those of a growing child.) Amino acids that need to be obtained from the diet are called essential amino acids. Nonessential amino acids are produced in the body. The pathways for the synthesis of nonessential amino acids are quite simple. Glutamate dehydrogenase catalyses the reductive amination of α-ketoglutarate to glutamate. A transamination reaction takes place in the synthesis of most amino acids. At this step, the chirality of the amino acid is established. Alanine and aspartate are synthesized by the transamination of pyruvate and oxaloacetate, respectively. Glutamine is synthesized from NH4+ and glutamate, and asparagine is synthesized similarly. Proline and arginine are derived from glutamate. Serine, formed from 3-phosphoglycerate, is the precursor of glycine and cysteine. Tyrosine is synthesized by the hydroxylation of phenylalanine, an essential amino acid. The pathways for the biosynthesis of essential amino acids are much more complex than those for the nonessential ones.

Tetrahydrofolate, a carrier of activated one-carbon units, plays an important role in the metabolism of amino acids and nucleotides. This coenzyme carries one-carbon units at three oxidation states, which are interconvertible: most reduced—methyl; intermediate—methylene; and most oxidized—formyl, formimino, and methenyl. The major donor of activated methyl groups is S-adenosylmethionine, which is synthesized by the transfer of an adenosyl group from ATP to the sulfur atom of methionine. S-Adenosylhomocysteine is formed when the activated methyl group is transferred to an acceptor. It is hydrolyzed to adenosine and homocysteine, the latter of which is then methylated to methionine to complete the activated methyl cycle.

Amino Acid Biosynthesis is Regulated by Feedback Inhibition

Most of the pathways of amino acid biosynthesis are regulated by feedback inhibition, in which the committed step is allosterically inhibited by the final product. Branched pathways require extensive interaction among the branches that includes both negative and positive regulation. The regulation of glutamine synthetase from E. coli is a striking demonstration of cumulative feedback inhibition and of control by a cascade of reversible covalent modifications.

Amino Acids are Precursors of Many Biomolecules

Amino acids are precursors of a variety of biomolecules. Glutathione (γ-Glu-Cys-Gly) serves as a sulfhydryl buffer and detoxifying agent. Glutathione peroxidase, a selenoenzyme, catalyses the reduction of hydrogen peroxide and organic peroxides by glutathione. Nitric oxide, a short-lived messenger, is formed from arginine. Porphyrins are synthesized from glycine and succinyl CoA, which condense to give δ-aminolevulinate. Two molecules of this intermediate become linked to form porphobilinogen. Four molecules of porphobilinogen combine to form a linear tetrapyrrole, which cyclizes to uroporphyrinogen III. Oxidation and side-chain modifications lead to the synthesis of protoporphyrin IX, which acquires an iron atom to form heme.

Non-protein Functions

In humans, non-protein amino acids also have important roles as metabolic intermediates, such as in the biosynthesis of the neurotransmitter gamma-aminobutyric acid. Many amino acids are used to synthesize other molecules, for example:

- Tryptophan is a precursor of the neurotransmitter serotonin.
- Tyrosine is a precursor of the neurotransmitter dopamine.
- Glycine is a precursor of porphyrins such as heme.
- Arginine is a precursor of nitric oxide.
- Ornithine and S-adenosylmethionine are precursors of polyamines.
- Aspartate, glycine and glutamine are precursors of nucleotides.
- Phenylalanine is a precursor of various phenylpropanoids which are important in plant metabolism.

However, not all of the functions of other abundant non-standard amino acids are known, for example taurine is a major amino acid in muscle and brain tissues, but although many functions have been proposed, its precise role in the body has not been determined.

Some non-standard amino acids are used as defences against herbivores in plants. For example canavanine is an analogue of arginine that is found in many legumes, and in particularly large amounts in *Canavalia gladiata* (sword bean). This amino acid protects the plants from predators such as insects and can cause illness in people if some types of legumes are eaten without processing. The non-protein amino acid mimosine is found in other species of legume, particularly *Leucaena leucocephala*. This compound is an analogue of tyrosine and can poison animals that graze on these plants.

Uses in Technology

Amino acids are used for a variety of applications in industry but their main use is as additives to animal feed. This is necessary since many of the bulk components of these feeds, such as soybeans, either have low levels or lack some of the essential amino acids: lysine, methionine, threonine, and tryptophan are most important in the production of these feeds. The food industry is also a major consumer of amino acids, particularly glutamic acid, which is used as a flavor enhancer, and Aspartame (aspartyl-phenylalanine-1-methyl ester) as a low-calorie artificial sweetener. The remaining production of amino acids is used in the synthesis of drugs and cosmetics.

Expanded Genetic Code

An expanded genetic code refers to an artificially modified genetic code in which one or more specific codons have been allocated to encode an amino acid which is not among the twenty/twentytwo found in nature.

In order to understand how expansion of the genetic code is achieved, the translational mechanism catalysed by ribosomes should be considered: the central parts are the transfer RNAs (tRNA) which are used as keys to decode a three nucleotide code on the RNA being translated into its equivalent amino acid. To do this the tRNA recognises a specific codon thanks to a complementary sequence called the anticodon on one of its loops. As a result generally there are one (or more) tRNAs for each codon (or group of degenerate codons). The encoding of a codon to its amino acid is a result of the aminoacyl tRNA synthetase which adds the aminoacyl group to its allocated tRNA.

However, the aminoacyl tRNA synthetase often recognises a specific motif irrespective of the anticodon, meaning that if the anticodon were to be mutated the encoding of that amino acid would change to a new codon. For successful translation of a novel amino acid, the codon to which the amino acid is reassigned must be free or unfavoured and the novel tRNA and synthetase set (called the orthogonal set when including the codon) must not crosstalk with the endogenous tRNA and synthetase sets, while still being functionally compatible with the ribosome and other components of the translation apparatus. The tRNA synthetase pair is taken from a distant organism, generally from a different domain, and the active site of the synthetase is modified to accept the non-natural amino acid.

The possibility of reassigning codons was realized by Normanly et al. in 1990 when a viable mutant strain of E. coli read through the amber codon. As a result the amber codon became the choice codon to be assigned a novel amino acid. Later, in the Schultz lab the tRNATyr/tyrosyl-tRNA synthetase (TyrRS) from Methanococcus jannaschii was used to introduce a tyrosine instead of STOP, the default value of the amber codon. As mentioned, this was possible because of the differences between the endogenous bacterial synthases and the orthologous archeal synthase which do not recognise each other.

Directed Evolution

This orthologous set can then be mutated and screened through directed evolution to accept a different, even novel, amino acid. Mutations to the plasmid containing the pair can be introduced by error-prone PCR or through degenerate primers for the synthetase's active site. Selection involves multiple rounds of a two-step process, where the plasmid is transferred into cells expressing chloramphenicol acetyl transferase with a premature amber codon.

In the presence of toxic chloramphenicol and the non-natural amino acid, the surviving cells will have overridden the amber codon using the orthogonal tRNA aminoacylated with either the standard amino acids or the non-natural one. To remove the former, the plasmid is inserted into cells with a barnase gene (toxic) with a premature amber codon but without the non-natural amino acid, removing all the orthogonal synthases which do not specifically recognize the non-natural amino acid.

In addition to the recoding of the tRNA to a different codon, they can be mutated to recognize a four base codon, allowing additional free coding options. The non natural amino acid, as a result, introduces diverse physicochemical and biological properties in order to be used as a tool to explore protein structure and function or to create novel or enhanced protein for practical purposes.

Diversity

The orthogonal pairs of synthase and tRNA which work for one organism may not work for another as the synthase may mis-aminoacylate endogenous tRNAs or the tRNA be mis-aminoacylated itself by an endogenous synthase. As a result the sets created to date differ between organisms.

Orthogonal Sets in E. Coli

- tRNA^{Tyr}-TyrRS pair from the archaeon *Methanococcus jannaschii*
- tRNA^{Lys}–LysRS pair from the archaeon *Pyrococcus horikoshii*
- tRNA^{Glu}–GluRS pair from *Methanosarcina mazei*
- leucyl-tRNA synthetase from *Methanobacterium thermoautotrophicum* and a mutant leucyl tRNA derived from *Halobacterium* sp.

Orthogonal Sets in Yeast

- tRNA^{Tyr}-TyrRS pair from *Escherichia coli*
- tRNA^{Leu}–LeuRS pair from *Escherichia coli*
- tRNA^{iMet} from human and GlnRS from *Escherichia coli*.

Orthogonal sets in Mammalian Cells

- tRNA^{Tyr}-TyrRS pair from *Bacillus stearothermophilus*
- modified tRNA^{Trp}-TrpRS pair from *Bacillus subtilis* trp
- tRNA^{Leu}–LeuRS pair from *Escherichia coli*.

Protein Studies

With an expanded genetic code, the unnatural amino acid can be genetically directed to any chosen site in the protein of interest. The high efficiency and fidelity allows a better control of the placement of the modification compared to modifying the protein post-translationally, which generally will target all amino acids of the same type, such as the thiol group of cysteine and the-amino group of lysine. Also, an expanded genetic code allows modifications to be carried out *in vivo*. The ability to site-specifically direct lab-synthesized chemical moieties into proteins allows many types of studies which would otherwise be extremely difficult.

- Probing Protein Structure and Function: by using amino acids with slightly different size such as o-Methyltyrosine or dansylalanine instead of tyrosine, and by inserting genetically-coded reporter moieties (colour-changing and/or spin-active) into selected protein sites, chemical information about the protein's structure and function can be measured.

- Identifying and Regulating Protein Activity: by using photocaged aminoacids, protein function can be "switched" on or off by illuminating the organism.

- Changing the mode of action of a protein: one can start with the gene for a protein which binds a certain sequence of DNA, and, by inserting a chemically active amino acid into the binding site, convert it to a protein which cuts the DNA, rather than binding it.

- Improving immunogenicity and overcoming self-tolerance: by replacing strategically-chosen tyrosines with *p*-nitro phenylalanine, a tolerated self-protein can be made immunogenic.

An example of the possible application for this method is the biomedical where "chemical warheads" can be added to protein which target specific cellular components.

Asymmetric Synthesis

Asymmetric synthesis, also called chiral synthesis, enantioselective synthesis or stereoselective synthesis, is organic synthesis that introduces one or more new and desired elements of chirality. This is important in the field of pharmaceuticals because the different enantiomers or diastereomers of a molecule often have different biological activity.

Approaches

There are three main approaches to asymmetric synthesis:

- chiral pool synthesis
- chiral auxiliaries
- asymmetric catalysis.

In practice, a mixture of all three is often used in order to maximize the advantages of each method.

Chirality must be introduced to the substance first. Then, it must be maintained. Care needs to be taken when planning the synthesis: The chirality might be removed by a chemical change that makes the substance isotropic. This process is called epimerization. For example, a S_N1 substitution reaction converts a molecule that is chiral by merit of non-planarity into a planar molecule, which has no handedness. (To visualise, draw the outlines of both of your hands on paper, and cut the images out. You can now superimpose the images, even if the hands themselves do not superimpose.) In a S_N2 substitution reaction, on the other hand, the chirality inverts, i.e., when you start with a right-handed mixture, you'll end up with left-handed one. (A visualization could be inverting an umbrella. The mechanism looks just the same.)

Chiral Pool Synthesis

Chiral pool synthesis is the easiest approach: A chiral starting material is manipulated through successive reactions using achiral reagents that retain its chirality to obtain the desired target molecule. This is especially attractive for target molecules having the similar chirality to a relatively inexpensive naturally occurring building-block such as a sugar or amino acid. However, the number of possible reactions the molecule can undergo is restricted, and tortuous synthetic routes may be required. Also, this approach requires a stoichiometric amount of the enantiopure starting material, which may be rather expensive if not occurring in nature, whereas chiral catalysis requires only a catalytic amount of chiral material.

Asymmetric Induction

What many strategies in chiral synthesis have in common is asymmetric induction. The aim is to make enantiomers into diastereomers, since diastereomers have different reactivity, but enantiomers do not. To make enantiomers into diastereomers, the reagents or the catalyst need to be incorporated with an enantiopure chiral centre. The reaction will now proceed differently for different enantiomers, because the transition state of the reaction can exist in two diastereomers with respect to the enantiopure centre, and these diastereomers react differently.

Asymmetric induction can also occur intramolecularly when given a chiral starting material. This chirality transfer can be exploited, especially when the goal is to make several consecutive chiral centres to give a specific enantiomer of a specific diastereomer. An aldol reaction, for example, is inherently diastereoselective; if the aldehyde is enantiopure, the resulting aldol adduct is diastereomerically and enantiomerically pure.

Chiral Auxiliary

One asymmetric induction strategy is the use of a chiral auxiliary, which forms an adduct to the starting materials and physically blocks the other trajectory for attack, leaving only the desired trajectory open. Assuming the chiral auxiliary is enantiopure, the different trajectories are not equivalent, but diastereomeric. The auxiliary shares problems similar to protecting groups; like protecting groups, auxiliaries require a reaction step to add and another to remove, increasing cost and decreasing yield.

Asymmetric Catalysis

The oldest asymmetric synthesis is the enantioselective decarboxylation of the malonic acid *2-ethyl-2-methylmalonic acid* mediated by brucine (forming the salt) as reported by Willy Marckwald in 1904:

Small amounts of chiral, enantiomerically pure (or enriched) catalysts promote reactions and lead to the formation of large amounts of enantiomerically pure or enriched products. Mostly, three different kinds of chiral catalysts are employed:

1. metal ligand complexes derived from chiral ligands
2. chiral organocatalysts
3. biocatalysts.

The first methods were pioneered by William S. Knowles and Ryôji Noyori (Nobel Prize in Chemistry 2001). Knowles in 1968 replaced the achiral triphenylphosphine ligands in Wilkinson's catalyst by the chiral phosphine ligands P(Ph)(Me)(Propyl), thus creating the first asymmetric catalyst.

This experimental catalyst was employed in an asymmetric hydrogenation with a modest 15% enantiomeric excess result. The methodology was ultimately used by him (while working for the Monsanto Company company) in an asymmetric hydrogenation step in the industrial production of L-DOPA. In the same year and independently, Noyori published his chiral ligand for a cyclopropanation reaction of styrene. In common with Knowles' findings, Noyori's results for the enantiomeric excess for this first-generation ligand was disappointingly low: 6%.

Examples of asymmetric catalysis include:

- BINAP, a chiral phosphine, used in combination with compounds of ruthenium or rhodium. These complexes catalyse the hydrogenation of functionalised alkenes well on only one face of the molecule. This process also developed by Ryôji Noyori is commercialized as the industrial synthesis of menthol using a chiral BINAP-rhodium complex.
- The other part of that Nobel prize concerned the Sharpless bishydroxylation
- Naproxen is synthesized with a chiral phosphine ligand in a hydrocyanation reaction
- asymmetric catalytic reduction and oxidation.

Biocatalysis & Organocatalysis

Biocatalysis makes use of enzymes to effect chemical reagents stereoselectively. Some small organic molecules can also be used to help accelerate the desired reaction; this method is known as organocatalysis. If the organic molecule is chiral, it may react preferentially with the substrate of a certain chirality.

Alternatives

Apart from asymmetric synthesis, racemic mixtures of compounds may be separated by various techniques in chiral resolution. Where the cost in time and money of making such racemic mixtures is low, or if both enantiomers may find use, this approach may remain cost-effective.

Biodegradable Plastic

Biodegradable plastics are plastics that will decompose in natural aerobic (composting) and anaerobic (landfill) environments. Biodegradation of plastics can be achieved by enabling microorganisms in the environment to metabolize the molecular structure of plastic films to produce an inert humus-like material that is less harmful to the environment. They may be composed of either bioplastics, which are plastics whose components are derived from renewable raw materials, or petroleum-based plastics which utilize an additive. The use of bio-active compounds compounded with swelling agents ensures that, when combined with heat and moisture, they expand the plastic's molecular structure and allow the bio-active compounds to metabolize and neutralize the plastic. Biodegradable plastics typically are produced in two forms: injection molded (solid, 3D shapes), typically in the form of disposable food service items, and films, typically organic fruit packaging and collection bags for leaves and grass trimmings, and agricultural mulch.

Scientific Definitions of Biodegradable Plastic

In the United States, the FTCFederal Trade Commission is the authoritative body for biodegradable standards. ASTM International defines appropriate testing methods to test for biodegradable plastic, both anaerobically and aerobically as well as in marine environments. The specific subcommittee responsibility for overseeing these standards falls on the Committee D20.96 on Environmentally Degradable Plastics and Biobased Products. The current ASTM standards are defined as standard specifications and standard test methods. Standard specifications create a pass or fail scenario whereas standard test methods identify the specific testing parameters for facilitating specific time frames and toxicity of biodegradable tests on plastics. Currently, there are three such ASTM standard specifications which mostly address biodegradable plastics in composting type environments, the ASTM D6400-04 Standard Specification for Compostable Plastics, ASTM D6868 - 03 Standard Specification for Biodegradable Plastics Used as Coatings on Paper and Other Compostable Substrates, and the ASTM D7081 - 05 Standard Specification for Non-Floating Biodegradable Plastics in the Marine Environment. The most accurate standard test method for anaerobic environments is the ASTM D5511 - 02 Standard Test Method for Determining Anaerobic Biodegradation of Plastic Materials Under High-Solids Anaerobic-Digestion Conditions. Another standard test method for testing in anaerobic environments is the ASTM D5526 - 94(2002) Standard Test Method for Determining Anaerobic Biodegradation of Plastic Materials Under Accelerated Landfill Conditions, this test has proven extremely difficult to perform. Both of these tests are used for the ISO DIS 15985 on determining anaerobic biodegradation of plastic materials.

Examples of Biodegradable Plastics

- While aromatic polyesters are almost totally resistant to microbial attack, most aliphatic polyesters are biodegradable due to their potentially hydrolysable ester bonds:
  - — Naturally Produced: Polyhydroxyalkanoates (PHAs) like the poly-3-hydroxybutyrate (PHB), polyhydroxyvalerate (PHV) and polyhydroxyhexanoate (PHH);
  - — Renewable Resource: Polylactic acid (PLA);
  - — Synthetic: Polybutylene succinate (PBS), polycaprolactone (PCL)...
- Polyanhydrides
- Polyvinyl alcohol
- Most of the starch derivatives
- Cellulose esters like cellulose acetate and nitrocellulose and their derivatives (celluloid).

Environmental Benefits of Biodegradable Plastics Depend upon Proper Disposal

Biodegradable plastics are not a panacea, however. Some critics claim that a potential environmental disadvantage of certified biodegradable plastics is that the carbon that is locked up in them is released into the atmosphere as a greenhouse gas. However, biodegradable plastics from natural materials, such as vegetable crop derivatives or animal products, sequester CO_2 during the phase when they're growing, only to release CO_2 when they're decomposing, so there is no net gain in carbon dioxide emissions.

However, certified biodegradable plastics require a specific environment of moisture and oxygen to biodegrade, conditions found in professionally managed composting facilities. There is much debate about the total carbon, fossil fuel and water usage in processing biodegradable plastics from natural materials and whether they are a negative impact to human food supply. Traditional plastics made from non-renewable fossil fuels lock up much of the carbon in the plastic as opposed to being utilized in the processing of the plastic.

The carbon is permanently trapped inside the plastic lattice, and is rarely recycled.

There is concern that another greenhouse gas, methane, might be released when any biodegradable material, including truly biodegradable plastics, degrades in an anaerobic (landfill) environment. Methane production from landfills is rarely captured or burned, but rather enter the atmosphere, where it is a potent greenhouse gas.

Methane production from specially managed landfill environments is captured and used for energy or burnt off to reduce the release of methane in the environment. Most landfills today capture the methane biogas for use in clean inexpensive energy. Of course, incinerating non-biodegradable plastics will release carbon dioxide as well. Disposing of biodegradable plastics made from natural materials in anaerobic (landfill) environments will result in the plastic lasting for hundred of years.

It is also possible that bacteria will eventually develop the ability to degrade plastics. This has already happened with nylon: two types of nylon eating bacteria, *Flavobacteria* and *Pseudomonas*, were found in 1975 to possess enzymes (nylonase) capable of breaking down nylon. While not a solution to the disposal problem, it is likely that bacteria will evolve the ability to use other synthetic plastics as well. In 2008, a 16-year-old boy reportedly isolated two plastic-consuming bacteria.

The latter possibility was in fact the subject of a cautionary novel by Kit Pedler and Gerry Davis (screenwriter), the creators of the Cybermen, re-using the plot of the first episode of their Doomwatch series. The novel, *Mutant 59: The Plastic Eater*, written in 1971, is the story of what could happen if a bacterium were to evolve—or be artificially cultured—to eat plastics, and be let loose in a major city.

Mechanisms

Materials such as a polyhydroxyalkanoate (PHA) biopolymer are completely compostable in an industrial compost facility. Polylactic acid (PLA) is another 100% compostable biopolymer which can fully degrade above 60C in an industrial composting facility. Fully biodegradable plastics are more expensive, partly because they are not widely enough produced to achieve large economies of scale.

Certain additives when added to conventional plastics attract the microbes to the molecular structure by allowing the hydrocarbons to be sensed once again by microbial colonies. When oil is in the ground, the microbes attach themselves onto the hydrocarbons consuming the oil and creating natural gas, 50% of which is methane gas. When the oil is cracked 4% is used for the plastic industry, if the plastic industry did not use this 4% the 4% would be considered waste and be thrown away or removed and dumped into a waste disposal facility, another 4% is used in the generation of your consumer product.

During this phase of cracking the organic compound which attracts the microbes to the molecular structure of the plastic is burnt out. The organic compound which is burnt out and other proprietary compounds which increase quorum sensing of the microbes and pH balance for the microbes are placed into the molecular structure of the plastic, to create a plastic product that can biodegrade 100 times faster than normal plastic.

Advantages and Disadvantages

Under proper conditions biodegradable plastics can degrade to the point where microorganisms can metabolise them.

Degradation of oil-based biodegradable plastics may release previously stored carbon as carbon dioxide. Starch-based bioplastics produced from sustainable farming methods can be almost carbon neutral but could have a damaging effect on soil, water usage and quality, and result in higher food prices.

There are concerns over "Oxo Biodegradable (OBD)" plastic bags. These are plastic bags which contain tiny amounts of metals such as cobalt, iron or manganese. They degrade in the presence of sunlight and oxygen, but there are concerns about the metals leftover and the time it takes for the plastics to degrade in certain circumstances.

Microbial consumption of polymers are available through addition of hydrophilic type additives onto the surface of the polymer chains. These types of additives are readily available and are used worldwide. The advantages of using these types of materials are heat stability, methane capturing and product performance.

Environmental Concerns; Benefits

Over 200 million tons of plastic are manufactured annually around the world, according to the Society of Plastics Engineers. Of those 200 million tons, 26 million are manufactured in the United States. The EPA reported in 2003 that only 5.8% of those 26 million tons of plastic waste are recycled, although this is increasing rapidly.

Much of the reason for disappointing plastics recycling goals is that conventional plastics are often commingled with organic wastes (food scraps, wet paper, and liquids), making it difficult and impractical to recycle the underlying polymer without expensive cleaning and sanitizing procedures.

On the other hand, composting of these mixed organics (food scraps, yard trimmings, and wet, non-recyclable paper) is a potential strategy for recovering large quantities of waste and dramatically increase community recycling goals. Food scraps and wet, non-recyclable paper comprises 50 million tons of municipal solid waste. Biodegradable plastics can replace the non-degradable plastics in these waste streams, making municipal composting a significant tool to divert large amounts of otherwise nonrecoverable waste from landfills. If even a small amount of conventional plastics were to be commingling with organic materials, the entire batch of organic waste is "contaminated" with small bits of plastic that spoil prime-quality compost humus. Composters, therefore, will not accept mixed organic waste streams unless they are completely devoid of nondegradable plastics. So, because of a relatively small quantity of nondegradable plastics, a significant waste disposal strategy is stalled.

However, proponents of biodegradable plastics argue that these materials offer a solution to this problem. Certified biodegradable plastics combine the utility of plastics (lightweight, resistance, relative low cost) with the ability to completely and fully biodegrade in a compost facility. Rather than worrying about recycling a relatively small quantity of commingled plastics, these proponents argue that certified biodegradable plastics can be readily commingled with other organic wastes, thereby enabling composting of a much larger position of nonrecoverable solid waste. Commercial composting for all mixed organics then becomes commercially viable and economically sustainable. More municipalities can divert significant quantities of waste from overburdened landfills since the entire waste stream is now biodegradable and therefore easier to process.

The use of biodegradable plastics, therefore, is seen as an enabler for the complete recovery of large quantities of municipal sold waste (via aerobic composting) that were are heretofore unrecoverable by other means except land filling or incineration.

Confusion over Proper Definition of Terms

Until recently there were few legal standards regarding marketing claims surrounding the use of the term 'biodegradable'. In 2007, the state of California passed regulation banning companies from claiming their products are biodegradable without proper scientific certification from a third-party laboratory. The Federal Court of Australia declared on March 30, 2009 that a director of a company that manufactured 'biodegradable' disposable diapers (who also approved the company's advertising) had been knowingly making false and misleading claims about biodegradability.

In June 2009, the Federal Trade Commission charged two companies with making unsupported marketing claims regarding biodegradability.

Energy Costs for Production

Various researchers have undertaken extensive life cycle assessments of biodegradable polymers to determine whether these materials are more energy efficient than polymers made by conventional fossil fuel-based means. Research done by Gerngross, *et al.* estimates that the fossil fuel energy required to produce a kilogram of polyhydroxyalkanoate (PHA) is 50.4 MJ/kg, which coincides with another estimate by Akiyama, *et al.*, who estimate a value between 50-59 MJ/kg. This information does not take into account the feedstock energy, which can be obtained from non-fossil fuel based methods. Polylactide (PLA) was estimated to have a fossil fuel energy cost of 54-56.7 from two sources, but recent developments in the commercial production of PLA by NatureWorks has eliminated some dependence fossil fuel based energy by supplanting it with wind power and biomass-driven strategies.

They report making a kilogram of PLA with only 27.2 MJ of fossil fuel-based energy and anticipate that this number will drop to 16.6 MJ/kg in their next generation plants. In contrast, polypropylene and high density polyethylene require 85.9 and 73.7 MJ/kg respectively, but these values include the embedded energy of the feedstock because it is based on fossil fuel. Gerngross reports a 2.65 total fossil fuel energy equivalent (FFE) required to produce a single kilogram of PHA, while polypropylene only requires 2.2 kg FFE. Gerngross assesses that the decision to proceed forward with any biodegradable polymer alternative will need to take into account the priorities of society with regard to energy, environment, and economic cost.

Furthermore, it is important to realize the youth of alternative technologies. Technology to produce PHA, for instance, is still in development today, and energy consumption can be further reduced by eliminating the fermentation step, or by utilizing food waste as feedstock. The use of alternative crops other than corn, such as sugar cane from Brazil, are expected to lower energy requirements-manufacturing of PHAs by fermentation in Brazil enjoys a favourable energy consumption scheme where bagasse is used as source of renewable energy.

Many biodegradable polymers that come from renewable resources (i.e., starch-based, PHA, PLA) also compete with food production, as the primary feedstock is currently corn. For the US to meet its current output of plastics production with BPs, it would require 1.62 square meters per kilogram produced. While this space requirement could be feasible, it is always important to consider how much impact this large scale production could have on food prices and the opportunity cost of using land in this fashion versus alternatives.

Biopolymer

Biopolymers are polymers produced by living organisms. Cellulose, starch and chitin, proteins and peptides, and DNA and RNA are all examples of biopolymers, in which the monomeric units, respectively, are sugars, amino acids, and nucleotides. Cellulose is both the most common biopolymer and the most common organic compound on Earth. About 33 percent of all plant matter is cellulose (the cellulose content of cotton is 90 percent and that of wood is 50 percent.

Some biopolymers are biodegradable. That is, they are broken down into CO_2 and water by microorganisms. In addition, some of these biodegradable biopolymers are compostable. That is, they can be put into an industrial composting process and will break down by 90% within 6 months. Biopolymers that do this can be marked with a 'compostable' symbol, under European Standard EN 13432 (2000). Packaging marked with this symbol can be put into industrial composting processes and will break down within 6 months (or less). An example of a compostable polymer is PLA film under 20 μm thick: films which are thicker than that do not qualify as compostable, even though they are biodegradable. A home composting logo may soon be established: this will enable consumers to dispose of packaging directly onto their own compost heap.

Biopolymers Versus Polymers

A major but defining difference between polymers and biopolymers can be found in their structures. Polymers, including biopolymers, are made of repetitive units called monomers. Biopolymers often have a well defined structure, though this is not a defining characteristic (example:ligno-cellulose): The exact chemical composition and the sequence in which these units are arranged is called the primary structure, in the case of proteins. Many biopolymers spontaneously fold into characteristic compact shapes, which determine their biological functions and depend in a complicated way on their primary structures. Structural biology is the study of the structural properties of the biopolymers. In contrast most synthetic polymers have much simpler and more random (or stochastic) structures.

This fact leads to a molecular mass distribution that is missing in biopolymers. In fact, as their synthesis is controlled by a template directed process in most in vivo systems all biopolymers of a type (say one specific protein) are all alike: they all contain the similar sequences and numbers of monomers and thus all have the same mass. This phenomenon is called monodispersity in contrast to the polydispersity encountered in synthetic polymers. As a result biopolymers have a polydispersity index of 1.

Conventions and Nomenclature

Polypeptides: The convention for a polypeptide is to list its constituent amino acid residues as they occur from the amino terminus to the carboxylic acid terminus. The amino acid residues are always joined by peptide bonds. Protein, though used colloquially to refer to any polypeptide, refers to larger or fully functional forms and can consist of several polypeptide chains as well as single chains. Proteins can also be modified to include non-peptide components, such as saccharide chains and lipids.

Nucleic Acids

The convention for a nucleic acid sequence is to list the nucleotides as they occur from the 5' end to the 3' end of the polymer chain, where 5' and 3' refer to the numbering of carbons around the ribose ring which participate in forming the phosphate diester linkages of the chain. Such a sequence is called the primary structure of the biopolymer.

Sugars

Sugar-based biopolymers are often difficult with regards to convention. Sugar polymers can be linear or branched are typically joined with glycosidic bonds. However, the exact placement of the linkage can vary and the orientation of the linking functional groups is also important, resulting in α- and β-glycosidic bonds with numbering definitive of the linking carbons' location in the ring. In addition, many saccharide units can undergo various chemical modification, such as amination, and can even form parts of other molecules, such as glycoproteins.

Structural Characterization

There are a number of biophysical techniques for determining sequence information. Protein sequence can be determined by Edman degradation, in which the N-terminal residues are hydrolyzed from the chain one at a time, derivatized, and then identified. Mass spectrometer techniques can also be used. Nucleic acid sequence can be determined using gel electrophoresis and capillary electrophoresis. Lastly, mechanical properties of these biopolymers can often be measured using optical tweezers or atomic force microscopy. Dual polarisation interferometry can be used to measure the conformational changes or self assembly of these materials when stimulated by pH, temperature, ionic strength or other binding partners.

Biopolymers as Materials

Some biopolymers- such as polylactic acid (PLA), naturally occurring zein, and poly-3-hydroxybutyrate can be used as plastics, replacing the need for polystyrene or polyethylene based plastics.

Some plastics are now referred to as being 'degradable', 'oxy-degradable' or 'UV-degradable'. This means that they break down when exposed to light or air, but these plastics are still primarily (as much as 98 per cent) oil-based and are not currently certified as 'biodegradable' under the European Union directive on Packaging and Packaging Waste (94/62/EC). Biopolymers, however, will break down and some are suitable for domestic composting.

Biopolymers as Packaging

Biopolymers (also called renewable polymers) are produced from biomass for use in the packaging industry. Biomass comes from crops such as sugar beet, potatoes or wheat: when used to produce biopolymers, these are classified as non food crops. These can be converted in the following pathways:

Sugar beet > Glyconic acid > Polyglonic acid.

Starch > (fermentation) > Lactic acid > Polylactic acid (PLA).

Biomass > (fermentation) > Bioethanol > Ethene > Polyethylene.

Many types of packaging can be made from biopolymers: food trays, blown starch pellets for shipping fragile goods, thin films for wrapping.

Biopolymers are Renewable, Sustainable, and can be Carbon Neutral

Biopolymers are renewable, because they are made from plant materials which can be grown year on year indefinitely. These plant materials come from agricultural non food crops. Therefore, the use of biopolymers would create a sustainable industry. In contrast, the feedstocks for polymers derived from petrochemicals will eventually run out. In addition, biopolymers have the potential to cut carbon emissions and reduce CO_2 quantities in the atmosphere: this is because the CO_2 released when they degrade can be reabsorbed by crops grown to replace them: this makes them close to carbon neutral.

Biopolymers are Biodegradable, and some are also Compostable

Some biopolymers are biodegradable: they are broken down into CO_2 and water by microorganisms. In addition, some of these biodegradable biopolymers are compostable: they can be put into an industrial composting process and will break down by 90% within 6 months. Biopolymers that do this can be marked with a 'compostable' symbol, under European Standard EN 13432 (2000). Packaging marked with this symbol can be put into industrial composting processes and will break down within 6 months (or less). An example of a compostable polymer is PLA film under 20μm thick: films which are thicker than that do not qualify as compostable, even though they are biodegradable. A home composting logo may soon be established: this will enable consumers to dispose of packaging directly onto their own compost heap. The standards for such a home composting logo have not yet been developed.

Reactions

As amino acids have both a primary amine group and a primary carboxyl group, these chemicals can undergo most of the reactions associated with these functional groups. These include nucleophilic addition, amide bond formation and imine formation for the amine group and esterification, amide bond formation and decarboxylation for the carboxylic acid group. The multiple side chains of amino acids can also undergo chemical reactions. The types of these reactions are determined by the groups on these side chains and are therefore different between the various types of amino acid.

Peptide Synthesis

In organic chemistry, peptide synthesis is the production of peptides, which are organic compounds in which multiple amino acids are linked via amide bonds which are also known as peptide bonds. The biological process of producing long peptides (proteins) is known as protein biosynthesis.

Chemistry

Peptides are synthesized by coupling the carboxyl group or C-terminus of one amino acid to the amino group or N-terminus of another. Due to the possibility of unintended reactions, protecting groups are usually necessary. Chemical peptide synthesis starts at the C-terminal end of the peptide and ends at the N-terminus. This is the opposite of protein biosynthesis, which starts at the N-terminal end.

Liquid-phase Synthesis

Liquid-phase peptide synthesis is a classical approach to peptide synthesis. It has been replaced in most labs by solid-phase synthesis. However, it retains usefulness in large-scale production of peptides for industrial purposes.

Solid-phase Synthesis

Solid-phase peptide synthesis (SPPS), pioneered by Robert Bruce Merrifield, resulted in a paradigm shift within the peptide synthesis community. It is now the accepted method for creating peptides and proteins in the lab in a synthetic manner. SPPS allows the synthesis of natural peptides which are difficult to express in bacteria, the incorporation of unnatural amino acids, peptide/protein backbone modification, and the synthesis of D-proteins, which consist of D-amino acids. Small solid beads, insoluble yet porous, are treated with functional units ('linkers') on which peptide chains can be built. The peptide will remain covalently attached to the bead until cleaved from it by a reagent such as anhydrous hydrogen fluoride or trifluoroacetic acid. The peptide is thus 'immobilized' on the solid-phase and can be retained during a filtration process, whereas liquid-phase reagents and by-products of synthesis are flushed away.

The general principle of SPPS is one of repeated cycles of coupling-wash-deprotection-wash. The free N-terminal amine of a solid-phase attached peptide is coupled to a single N-protected amino acid unit. This unit is then deprotected, revealing a new N-terminal amine to which a further amino acid may be attached. The superiority of this technique partially lies in the ability to perform wash cycles after each reaction, removing excess reagent with all of the growing peptide of interest remaining covalently attached to the insoluble resin.

The overwhelmingly important consideration is to generate extremely high yield in each step. For example, if each coupling step were to have 99% yield, a 26-amino acid peptide would be synthesized in 77% final yield (assuming 100% yield in each deprotection); if each step were 95%, it would be synthesized in 25% yield. Thus each amino acid is added in major excess (2~10x) and coupling amino acids together is highly optimized by a series of well-characterized agents. There are two majorly used forms of SPPS — Fmoc and Boc. Unlike ribosome protein synthesis, solid-phase peptide synthesis proceeds in a C-terminal to N-terminal fashion. The N-termini of amino acid monomers is protected by these two groups and added onto a deprotected amino acid chain. Automated synthesizers are available for both techniques, though many research groups continue to perform SPPS manually.

SPPS is limited by yields, and typically peptides and proteins in the range of 70 amino acids are pushing the limits of synthetic accessibility. Synthetic difficulty also is sequence dependent; typically amyloid peptides and proteins are difficult to make. Longer lengths can be accessed by using native chemical ligation to couple two peptides together with quantitative yields. Since its introduction over 40 years ago, SPPS has been significantly optimized. First, the resins themselves have been optimized. Furthermore, the 'linkers' between the C-terminal amino acid and polystyrene resin have improved attachment and cleavage to the point of mostly quantitative yields. The evolution of side chain protecting groups has limited the frequency of unwanted side reactions. In addition, the evolution of new activating groups on the carboxyl group of the incoming amino acid have improved coupling and decreased epimerization. Finally, the process itself has been optimized.

In Merrifield's initial report, the deprotection of the α-amino group resulted in the formation of a peptide-resin salt, which required neutralization with base prior to coupling. The time between neutralization of the amino group and coupling of the next amino acid allowed for aggregation of peptides, primarily through the formation of secondary structures, and adversely affected coupling. The Kent group showed that concomitant neutralization of the α-amino group and coupling of the next amino acid led to improved coupling. Each of these improvements has helped SPPS become the robust technique that it is today.

Fmoc Solid-phase Peptide Synthesis

The capacity for anhydrous hydrogen fluoride to degrade proteins during the final cleavage conditions led to a new α-amino protecting group based on 9-fluorenylmethyloxycarbonyl (Fmoc). The Fmoc method allows for a milder deprotection scheme. This method utilizes a base, usually piperidine (20-50%) in DMF in order to remove the Fmoc group to expose the α-amino group for reaction with an incoming activated amino acid. Unlike the acid used to deprotect the α-amino group in Boc methods, Fmoc SPPS uses a base, and thus the exposed amine is neutral. Therefore, no neutralization of the peptide-resin is required, but the lack of electrostatic repulsions between the peptides can lead to increased aggregation.

Along with the development of Fmoc SPPS, different resins have also been created to be removed by TFA. Similar to the Boc strategy, two primary resins are used, based on whether a C-terminal carboxylic acid or amide is desired. The Wang resin is the most commonly used resin for peptides with C-terminal carboxylic acids. If a C-terminal amide is desired, the Rink amide resin is used.

Semipermanent protecting groups are t-butyl based, and final cleavage of the protein from the resin and removal of permanent protecting groups is performed with TFA in the presence of scavengers. Thus, the Fmoc method is orthogonal in two directions: deprotection of the α-amino group and final cleavage from the resin occur by independent mechanisms. The resulting final product is a TFA salt, which is more difficult to solubilize than the fluoride salts generated in Boc SPPS. This method is thus milder than the Boc method because the deprotection/cleavage-from-resin steps occur with different conditions rather than with different reaction rates.

t-Boc Solid-phase Peptide Synthesis

The original method for the synthesis of proteins relied on tert-butoxycarbonyl (Boc) to temporarily protect the α-amino group. In this method, the Boc group is covalently bound to the amino group to suppress its nucleophilicity. The C-terminal amino acid covalently linked to the resin through a linker. Next, the Boc group is removed with acid, such as trifluoroacetic acid (TFA). This forms a positively-charged amino group, which is simultaneously neutralized and coupled to the incoming activated amino acid. Reactions are driven to completion by the use of excess (two- to four-fold) activated amino acid. After each deprotection and coupling step, a wash with N,N-dimethylformamide (DMF) is performed to remove excess reagents, allowing for high yields (~99%) during each cycle.

Importantly, improvements to the resin have enhanced its ability to withstand the repeated use of TFA during the deprotection step. Furthermore, different resins allow for different functional groups at the C-terminus. The oxymethylphenylacetamidomethyl (PAM) resin results in the conventional C-terminal carboxylic acid. On the other hand, the paramethylbenzhydrylamine (pMBHA) resin yields a C-terminal amide, which is useful in mimicking the interior of a protein.

Permanent side chain protecting groups are typically benzyl or benzyl-based groups. Final removal of the peptide from the linkage occurs simultaneously with side-chain deprotection with anhydrous hydrogen fluoride via hydrolytic cleavage. The final product is a fluoride salt which is relatively easy to solubilize. Importantly, scavengers such as cresol are added to the HF in order to prevent reactive t-butyl cations from generating undesired products. In fact, the use of harsh hydrogen fluoride may degrade some peptides, which was the premise for the development of a milder, base-labile method of SPPS—namely, the Fmoc method.

Some researchers prefer Boc SPPS for complex syntheses. In addition, when synthesizing nonnatural peptide analogs which are base-sensitive (such as depsipeptides), the t-Boc protecting group is necessary. This is because Fmoc SPPS uses base to protect the α-amino group.

Comparison of Boc and Fmoc Solid-Phase Peptide Synthesis

Both the Fmoc and Boc methods offer advantages and disadvantages. The selection of one technique over another is thus made on a case-by-case basis.

	Boc	*Fmoc*
Requires special equipment	Yes	No
Cost of reagents	Lower	Higher
Solubility of peptides	Higher	Lower
Purity of hydrophobic peptides	High	May be lower
Problems with aggregation	Less frequently	More frequently
Synthesis time	~20 min/amino acid	~20-60 min/amino acid
Final deprotection	HF	TFA
Safety	Potentially dangerous	Relatively safe
Orthogonal	N	Yes

Boc SPPS uses special equipment to handle the final cleavage and deprotection step, which requires anhydrous hydrogen fluoride. Because the final cleavage of the peptide with Fmoc SPPS uses TFA, this special equipment is not necessary. The solubility of peptides generated by Boc SPPS is generally higher than those generated with the Fmoc method, because fluoride salts are higher in solubility than TFA salts. Next, problems with aggregation are generally more of an issue with Fmoc SPPS. This is primarily because the removal of a Boc group with TFA yields a positively-charged α-amino group, whereas the removal of an Fmoc group yields a neutral α-amino group. The steric hindrance of the positively charged α-amino group limits the formation of secondary structure on the resin. Finally, the Fmoc method is considered orthogonal, since α-amino group deprotection is with base, while final cleavage from the resin is with acid. The Boc method utilizes acid for both deprotection and cleavage from the resin. Based on this comparison, one sees that both methods possess advantages and disadvantages. Thus, several factors help to decide which method may be preferable.

BOP SPPS

The use of BOP reagent was first described by Castro et al. in 1975.

Solid Supports

The name solid support implies that reactions are carried out on the surface of the support, but this is not the case. Reactions also occur within these particles, and thus the term "solid support" better describes the insolubility of the polymer. The physical properties of the solid support, and the applications to which it can be utilized, vary with the material from which the support is constructed, the amount of cross-linking, as well as the linker and handle being used. Most scientists in the field believe that supports should have the minimum amount of cross-linking to confer stability. This should result in a well-solvated system where solid-phase peptide synthesis can be carried out. Nonetheless, the characteristics of an efficient solid support include:

1. It must be physically stable and permit the rapid filtration of liquids, such as excess reagents

2. It must be inert to all reagents and solvents used during SPPS

3. It must swell extensively in the solvents used to allow for penetration of the reagents

4. It must allow for the attachment of the first amino acid.

There are four primary types of solid supports:

1. Gel-type supports: These are highly solvated polymers with an equal distribution of functional groups. This type of support is the most common, and includes:

   — Polystyrene: Styrene cross-linked with 1-2% divinylbenzene

   — Polyacrylamide: A hydrophilic alternative to polystyrene

   — Polyethylene glycol (PEG): PEG-Polystyrene (PEG-PS) is more stable than polystyrene and spaces the site of synthesis from the polymer backbone

— PEG-based supports: Composed of a PEG-polypropylene glycol network or PEG with polyamide or polystyrene

2. Surface-type supports: Many materials have been developed for surface functionalization, including controlled pore glass, cellulose fibres, and highly cross-linked polystyrene.

3. Composites: Gel-type polymers supported by rigid matrices.

Polystyrene Resin

Polystyrene resin is a versatile resin and it is quite useful in multi-well, automated peptide synthesis, due to its minimal swelling in dichloromethane. The initial support used by R. Bruce Merrifield was polysytrene cross-linked with 2% divinylbenzene. This support is sometimes referred to as the 'Merrifield resin.' This produces a hydrophobic bead that is solvated by nonpolar solvent such as dichloromethane. Since then, new resins have been developed with the following advantages:

1. Enhanced swelling or rigidity (a property of mechanical strength)

2. Chemical inertness.

Highly cross-linked (50%) polystyrene has been developed that possesses the features of increased mechanical stability, better filtration of reagents and solvents, and rapid reaction kinetics.

Polyamide Resin

Polyamide resin is also a useful and versatile resin. It seems to swell much more than polystyrene, in which case it may not be suitable for some automated synthesizers, if the wells are too small.

PEG Hybride Polystyrene Resin

An example of this type of resin is the Tentagel resin. The base resin is polystyrene onto which is attached long chains (Mw ca. 3000 Da) of polyethylene glycol (PEG; also known as polyethylene oxide). Synthesis is caried out on the distal end of the PEG spacer making it suited for long and difficult peptides. In addition it is also attractive for the synthesis of combinatorial peptide libraries and on resin screening experiments. It does not expand much during synthesis making it a preferred resin for robotic peptide synthesis.

PEG Based Resin

ChemMatrix(R) is a new type of resin which is based on PEG that is crosslinked. ChemMatrix(R) has claimed a high chemical and thermal stability (is compatible with Microwave synthesis) and has shown higher degrees of swellings in acetonitrile, dichloromethane, DMF, Nmethylpyrrolidone, TFA and water compared to the polystyrene based resins. ChemMatrix has shown significant improvements to the synthesis of hydrophobic sequences. ChemMatrix is recommended for the synthesis of difficult and long peptides.

Protecting Groups

Due to amino acid excesses used to ensure complete coupling during each synthesis step, polymerization of amino acids is common in reactions where each amino acid is not protected. In order to prevent this polymerization, protecting groups are used. This adds additional deprotection phases to the synthesis reaction, creating a repeating design flow as follows:

- Protecting group is removed from trailing amino acids in a deprotection reaction
- Deprotection reagents washed away to provide clean coupling environment
- Protected amino acids dissolved in a solvent such as dimethylformamide (DMF) are combined with coupling reagents are pumped through the synthesis column
- Coupling reagents washed away to provide clean deprotection environment.

Currently, two protecting groups (t-Boc, Fmoc) are commonly used in solid-phase peptide synthesis. Their lability is caused by the carbamate group which readily releases CO_2 for an irreversible decoupling step.

Bibliography

Alka Pareek: *Environment and Nutritional Disorders*, Aavishkar Pub, Delhi, 2003.

Arthur, W. Gilbert, Mortier F. Barrus and Daniel Dean: *Growing and Breeding of Potatoes*, Asiatic Pub, Delhi, 2006.

Arun Rastogi: *An Introduction to Practical Biochemistry*, Anmol, Delhi, 2010.

Ausubel, F.M.: *Current Protocols in Molecular Biology*, New York: John Wiley and Sons, 1989.

Aw, S. E.: *Chemical Evolution*, Singapore, University Education Press, 1976.

Banerjee, S.: *Cell Biochemistry*, Dominant Pub, Delhi, 2009.

Bhatnagar, Vasudev: *Cell Science and Technology*, Campus Books, Delhi, 2009.

Brandwein, P.F.: *Sourcebook for the Biological Sciences*, San Diego: Harcourt Brace Jovanovich, 1986.

Broach, J.R.: *The Molecular Biology of the Yeast*, Cold Spring Harbor, Cold Spring Harbor Laboratory, 1981.

Brock, H.: *History of Chemistry*, New York: Norton, 1992.

Cahill, Lisa: *Genetics, Theology, and Ethics: An Interdisciplinary Conversation*, New York: Crossroad, 2005.

Chaudhary, Vikas: *Entomology and Pest Management*, Navyug, Delhi, 2008.

Clark, J.M.: *Experimental Biochemistry*, New York, W.H. Freeman and Company, 1977.

Collymore L.: *Fruit Production in Barbados*, Port of Spain, Trinidad and Tobago, 1996.

Currah L. and Proctor F. J.: *Onions in Tropical Regions*, Kent, Natural Resources Institute, 1990.

Daphne C. Elliott: *Biochemistry and Molecular Biology*, Oxford University Press, Delhi, 2005.

David Sadava: *Plants, Genes and Crop Biotechnology*, Sudbury MA, Jones and Barlett Publishers, 2003.

Dudley, E.: *The Critical Villager: Beyond Community Participation*, London, Routledge, 1993.

Featherly H. I.: *Taxonomic Terminology of the Higher Plants*, USA, Iowa State College Press, 1954.

Ferentinos L.: *Proceeding of the Sustainable Taro Culture for the Pacific Conference*, Honolulu, HITAHR, 1993.

Fransman M, Junne G, Roobeek A: *The Biotechnology Revolution?*, Oxford, Blackwell, 1995.

Friedberg, E.C.: *DNA Repair*, New York, WH Freeman and Company, 1985.

Fumento, Michael: *Bioevolution: How Biotechnology is Changing Our World*, San Francisco, Encounter Books, 2003.

Ganguly, Smriti: *Biochemistry of Biomolecules,* Pearl Books, Delhi, 2007.

Geis, I.: *Chemistry, Matter, and the Universe*, Menlo Park, Ca., W.A. Benjamin, 1976.

Graham L. Patrick: *An Introduction to Medicinal Chemistry*, Oxford University Press, Delhi, 2009.

Haber, L. F.: *The Chemical Industry: 1900-1930, International Growth and Technological Change*, Oxford: Claredon Press, 1971.

Hardy B.: *Biology and Agronomy of Forage Arachis*, Cali, International Centre for Tropical Agriculture, 1994.

Harsh Bhaskar: *Basic Facts on Biochemistry*, Campus Books, Delhi, 2010.

Hayes, Williams: *American Chemical Industry: Background & Beginnings*, New York: D. Van Nostrand Company: 1954.

Iqbal, S.A. and M. Satake: *An Introduction to Analytical Chemistry*, Discovery, Delhi, 1999.

Jagbir Sharma: *Advanced Environment Chemistry*, RBSA Pub, Delhi, 2006.

Jagdish Chander and Anil Kumar: *A Comprehensive Text Book of Applied Chemistry*, Abhishek Pub, Delhi, 2009.

Kurzweil, Ray: *The Age of Spiritual Machines*, New York, Penguin Books, 1999.

Larry V. McIntire: *Biotechnology: Science, Engineering, and Ethical Challenges for the Twenty-first Century*, Washington, DC: Joseph Henry Press, 1996.

M. Prakash: *Cell Physiology*, Discovery, Delhi, 2010.

M. Satake and Y. Mido: *An Introduction to Nuclear Chemistry*, Discovery, Delhi, 1995.

M.L. Jangir: *Cell Biology: Fundamentals and Applications*, Agrobios, Delhi, 2009.

Muneesh Kainth: *Chordate Embryology*, Dominant, Delhi, 2003.

N. Sarath Chandra Bose: *Biochemistry: A Practical Manual*, Pharma Med Press, Delhi, 2010.

Nisha Khalsa: *Essentials of Biochemistry*, Aavishkar Pub, Delhi, 2008.

Nitin Suri: *Molecular Biology and Biochemistry*, Oxford Book Company, Delhi, 2010.

Nobel, P. S.: *Physicochemical and Environmental Plant Physiology*, Academic Press, San Diego, 1999.

Old, R.W.: *Principles of Gene Manipulation*, London, Blackwell Scientific Publications, 1989.

Oldham P.: *Cost of Production of Major Tree Crops in Dominica*, Roseau, Ministry of Agriculture, 1991.

Paul M. Althouse: *Introduction to Agricultural Biochemistry*, Biotech Books, Delhi, 2005.

Pemberton, R. W.: *Predictable Risk to Native Plants in Weed Biological Control*, Oecologia, 2000.

Pooja Bhagwan: *A Handbook of Inorganic Chemistry: Nuclear, Atomic, Aqueous, Alkali and Alkaline*, ISPA, Delhi, 2005.

Qystein V. Sjaastad: *Physiology of Domestic Animals*, International Book Distributing Co., Delhi, 2005.

Raghunath Narvekar: *Handbook of Biochemistry*, Adhyayan, Delhi, 2008.

Ragone D.: *Breadfruit: Artocarpus Altilis (Parkinson) Fosberg*, Rome, International Plant Genetic Resources Institute, 1997.

Richard Myers: *Basics of Chemistry*, Atlantic, Delhi, 2007.

Rifkin, Jeremy: *The Biotech Century*, New York, Penguin Putnam, 1998

Rutherford Lyn.: *A Gourmet s Book of Mushrooms & Truffles*, Sydney, Golden Press Pvt. Ltd., 1991.

S. Banerjee: *Cell and Resource Biology*, Dominant, Delhi, 2004.

Sharma, Pradeep: *Biochemistry and Organisation of Cells*, RBSA Pub, Delhi, 2006.

Smriti Ganguly: *Biochemistry of Biomolecules*, Pearl Books, Delhi, 2007.

Stewart Truswell: *Essentials of Human Nutrition*, Oxford University Press, Delhi, 2008.

Swarnim, K.: *A Textbook of Biochemistry and Microbiology*, Surendra Pub, Delhi, 2010.

Tawde, A. B.: *Propagation and Rootstocks of Mango*, New Delhi, Malhotra, 1993.

Urton, Gary: *The Social Life of Numbers*, Austin, University of Texas Press, 1997.

Van Antwerpen, F.J.: *The Origins of Chemical Engineering, History of Chemical Engineering*, Washington D.C.: ACS, 1980.

Vanangamudi, K.: *Principles and Methods of Plant Breeding*, International Book, Delhi, 2005.

Vasudev Bhatnagar: *Cell Science and Technology*, Campus Books, Delhi, 2009.

Whealy K.: *The Garden Seed Inventory*, Decorah, Seed Saver Publications, 1988.

White, G.F.: *Natural Hazards: Local, National, Global*, Oxford University Press, New York, 1974.

William Alec: *The Chemical Industry*, London: Longman Group Limited, 1971.

Index

A

Agar Gel, 40.
Anaerobic Biodegradation, 188, 258.

B

Basic Reproduction Number, 90, 92, 93, 95.
Bioeconomics, 41, 42.
Biological Engineering, 184, 185, 186.
Biopharmaceutical, 108, 109, 111.
Biopolitics, 42, 43.
Biotechnology Regulations, 192.

C

Catholicism, 13, 146.
Cell Culture, 22, 23, 24, 25, 27, 29, 37.
Cellular Model, 55.
Chimeric Plasmids, 16.
Comparative Genomics, 53, 54.
Controversial Cases, 170.
Critical Assessment, 57, 66, 83, 107.

D

De Novo Protein, 63, 84.
Deamidation, 212, 213.
DOCKing, 108.
Docking, 74, 82, 83, 100, 101, 103, 104, 105, 107, 108.

E

Ecosystem Model, 88, 89.
Endorsement, 6.

Epidemic Model, 89, 97, 99.
Established Human Cell, 36.

F

Fragment Assembly, 68.
Fungal Biodegradation, 191.

G

Gene Targeting, 43, 46, 47.
Gene Trapping, 47.
Genetically Modified Food, 43, 51, 52, 114, 158, 160, 174, 176, 182.
Genetics, 1, 31, 47, 55, 94, 110, 128, 129, 134, 170, 171, 184, 194, 195, 202, 204, 225, 239.
Glycosylation, 211, 212, 239.

H

Health Risks, 137, 159, 173, 178.
Homology Modelling, 63, 64, 65, 66, 67, 68, 69, 72, 73, 74, 75, 77, 82, 85, 105.
Human Biological Systems, 86.
Human Blood Clotting Factors, 20.
Human Cloning, 14, 138, 143, 144, 145, 146, 147, 148, 149, 150, 151.
Human Genome Project, 63, 117, 119, 121, 122, 125, 127, 129.
Human Insulin, 17, 19, 45.
Human Potential, 6.
Humanity, 6, 147, 187.
Hydroxylation, 213.

I

Intellectual Property, 52, 122, 153, 163, 167, 177.
Isomerism, 208.

L

Loop Modelling, 65, 69, 70.

M

Machine Learning, 58, 59, 62, 71.
Macromolecular Docking, 100.
Maintaining Cells, 23.
Mathematical Modelling, 42, 97.
Media Changes, 25, 41.
Methodology, 23, 39, 82, 225, 256.
Microbial Biodegradation, 186.
Microbial Phylogeny, 201.
Microbiological Culture, 37.
Model Assessment, 66, 71.
Modelling Biological Systems, 54.
Molecular Clock, 193, 196, 197, 198, 199, 200, 202, 203, 236.
Molecular Cloning, 32, 46, 130, 133, 134.
Molecular Farming, 112, 113, 114.
Molecular Medicine, 204.
Molecular Phylogenetics, 193.
Molecular Wars, 196.

N

Nuclear Transfer, 1, 2, 134, 136, 140, 141, 142.

O

Oil Biodegradation, 190.
Organ Culture, 22, 39, 40.
Organism Cloning, 139.

P

Pentecostalism, 13.
Pharmaceutical Manufacturing, 19.

Pharming, 44, 110, 111, 112, 114, 115.
Phylogenetic Profiling, 80.
Plasma Clot, 40, 41.
Protein Primary Structure, 209, 220.
Protein Threading Software, 78, 82.
Proteinogenic Amino Acid, 206.

R

RNA Transfection, 29, 30, 31.

S

Scoring Functions, 64, 100, 104, 105, 106, 107.
Segment Matching, 69.
Somatic Cell, 1, 2, 3, 4, 133, 134, 135, 136, 140, 141, 143.
Somatic Cell Nuclear Transfer, 1, 2, 134, 136, 140, 141.
Stem Cell Controversy, 3.
Stem Cell Research, 1, 2, 3, 5, 6, 7, 9, 11, 12, 13, 44, 134, 135, 136, 137, 148.

T

Terminology, 26, 29, 98.
Tissue Culture, 22, 23, 24, 25, 37, 39, 40, 48, 134.
Tissue Engineering, 17, 18, 37.
Transfection, 16, 21, 26, 27, 28, 29, 30, 31, 32, 36, 131.
Transformation, 21, 26, 31, 32, 33, 34, 35, 36, 131, 132.

U

Unicellular Organisms, 134.

V

Vector Transmission, 99.

W

Waste Biotreatment, 190.

□□□